NEXUS NETWORK JOURNA

T0214207

Aims and Scope

Founded in 1999, the *Nexus Network Journal* (NNJ) is a peer-reviewed journal for researchers, professionals and students engaged in the study of the application of mathematical principles to architectural design. Its goal is to present the broadest possible consideration of all aspects of the relationships between architecture and mathematics, including landscape architecture and urban design.

Editorial Office

Editor-in-Chief
Kim Williams
Corso Regina Margherita, 72
10153 Turin (Torino), Italy
E-mail: kwb@kimwilliamsbooks.com

Contributing Editors
The Geometer's Angle
Rachel Fletcher
113 Division St.
Great Barrington, MA 01230, USA
E-mail: rfletch@bcn.net

Book Reviews
Sylvie Duvernoy
Via Benozzo Gozzoli, 26
50124 Firenze, Italy
E-mail: syld@kimwilliamsbooks.com

Corresponding Editors
Alessandra Capanna
Via della Bufalotta 67
00139 Roma Italy
E-mail: alessandra.capanna@uniroma1.it

Tomás García Salgado
Palacio de Versalles # 200
Col. Lomas Reforma, c.p. 11930
México D.F., Mexico
E-mail: tgsalgado@perspectivegeometry.com

Robert Kirkbride
studio 'patafisico
12 West 29 #2
New York, NY 10001, USA
E-mail: kirkbrir@newschool.edu

Andrew I-Kang Li
Initia Senju Akebonocho 1313
Senju Akebonocho 40-1
Adachi-ku
Tokyo 120-0023 Japan
E-mail: i@andrew.li

Michael J. Ostwald
School of Architecture and Built Environment
Faculty of Engineering and Built Environment
University of Newcastle
New South Wales, Australia 2308
E-mail: michael.ostwald@newcastle.edu.au

Cover
View of the city of Porto
Photo ® Maycon Sedrez
Reproduced by permission

Vera S
The M
José M
E-mail: vspinaue@iiuci...

Igor Verner
The Department of Education in Technology and Science
Technion - Israel Institute of Technology
Haifa 32000, Israel
E-mail: ttrigor@techunix.technion.ac.il

Stephen R. Wassell
Department of Mathematical Sciences
Sweet Briar College, Sweet Briar, Virginia 24595, USA
E-mail: wassell@sbc.edu

João Pedro Xavier
Faculdade de Arquitectura da Universidade do Porto
Rua do Gólgota 215, 4150-755 Porto, Portugal
E-mail: jpx@arq.up.pt

Instructions for Authors

Authorship
Submission of a manuscript implies:
- that the work described has not been published before;
- that it is not under consideration for publication elsewhere;
- that its publication has been approved by all coauthors, if any, as well as by the responsible authorities at the institute where the work has been carried out;
- that, if and when the manuscript is accepted for publication, the authors agree to automatically transfer the copyright to the publisher; and
- that the manuscript will not be published elsewhere in any language without the consent of the copyright holder.

Exceptions of the above have to be discussed before the manuscript is processed. The manuscript should be written in English.

Submission of the Manuscript

Material should be sent to Kim Williams
via e-mail to: kwb@kimwilliamsbooks.com
or via regular mail to: Kim Williams Books,
Corso Regina Margherita, 72,
10153 Turin (Torino), Italy

Please include a cover sheet with name of author(s), title or profession (if applicable), physical address, e-mail address, abstract, and key word list.

Contributions will be accepted for consideration to the following sections in the journal: research articles, didactics, viewpoints, book reviews, conference and exhibits reports.

Final PDF files
Authors receive a pdf file of their contribution in its final form. Orders for additional printed reprints must be placed with the Publisher when returning the corrected proofs. Delayed reprint orders are treated as special orders, for which charges are appreciably higher. Reprints are not to be sold.

Articles will be freely accessible on our online platform SpringerLink two years after the year of publication.

Nexus Network Journal

SHAPE AND SHAPE GRAMMARS

Lionel March, Guest Editor

VOLUME 13, NUMBER 1
Spring 2011

KIM WILLIAMS BOOKS

Nexus Network Journal
Vol. 13
No. 1
Pp. 1-254
ISSN 1590-5896

CONTENTS

Kim Williams

Kim Williams Books
Corso Regina Margherita, 72
10153 Turin (Torino) ITALY
kwb@kimwilliamsbooks.com

João Nuno Tavares

Centro de Matemática da
Universidade do Porto (CMUP)
Departamentos de Matemática
Rua do Campo Alegre, 687
4169-007 Porto PORTUGAL
jntavar@fc.up.pt

João Pedro Xavier

Faculdade de Arquitectura da
Universidade do Porto (FAUP)
Via Panorâmica S/N
4150-755 Porto PORTUGAL
jpx@arq.up.pt

Conference Report

Nexus 2010: Relationships between Architecture and Mathematics

Abstract. The eighth edition of the international, interdisciplinary Nexus conference on architecture and mathematics took place from Sunday 13 June through Tuesday 15 June 2010, in Porto, Portugal.

Introduction

The 2010 edition of the international, interdisciplinary conference on architecture and mathematics took place from Sunday 13 June through Tuesday 15 June, in Porto (Portugal). The conference was directed by Kim Williams, founder of the Nexus conferences and editor in chief of the Nexus Network Journal, Prof. João Nuno Tavares of the Centro de Matemática da Universidade do Porto, and João Pedro Xavier of the Faculdade de Arquitectura da Universidade do Porto (FAUP). The conference presentations were given in the Auditório Fernando Távora of the FAUP, designed by Álvaro Siza.

This year's conference offered a program that was rich and varied. Each of the three days of the meeting featured morning and afternoon sessions dedicated to specific themes, and closed with the presentation of a keynote speaker. One of the new developments of the Nexus conference is that the papers presented will no longer be published in a conference proceedings book, but will be published in the *Nexus Network Journal*, where they will be more widely available for consultation and will enjoy a greater distribution. The papers from the 2010 conference are all scheduled for publication in the 3 issues of vol. 13 (2011), grouped according to theme as they were at the conference.

The morning session of Sunday 11 June was entitled "From Mediaeval Stonecutting to Projective Geometry" and was moderated by José Calvo Lopez of the Universidad Politécnica de Cartagena. Four presentations were given by Giuseppe Fallacara and Luc Tamboréro, Miguel Alonso, Elena Pliego and Alberto Sanjurjo, Snezana Lawrence, and Enrique Rabasa. These papers are scheduled for publication in the *Nexus Network Journal* vol. 13 no. 3 (Winter 2011). The afternoon session of 11 June was dedicated to a round table presentation entitled "Generative Architecture Codeness", moderated by Celestino Soddu of the Politecnico di Milano, with the participation of Marie-Pascale Corcuff, Philip van Loocke, Enrica Colabella, and Gabriel Maldonado.

DOI 10.1007/s00004-011-0065-5; *published online* 9 March 2011

The opening session of Nexus 2010 in Porto

U.PORTO

FACULDADE DE ARQUITECTURA
UNIVERSIDADE DO PORTO

CMUP

Centro de **Matemática**
Universidade do Porto

FCT

Fundação para a Ciência e a Tecnologia
MINISTÉRIO DA CIÊNCIA, TECNOLOGIA E ENSINO SUPERIOR

 FUNDAÇÃO CALOUSTE GULBENKIAN

Sponsors of Nexus 2010 in Porto

The morning session of Monday 12 June was entitled "Architecture, Systems Research and Computational Sciences", and was moderated by Gonçalo Furtado of the FAUP. Speakers were Pau Solá-Morales, Lora Dikova, Suzanne Strum, and Maycon Sedrez. These papers are also scheduled for publication in the *Nexus Network Journal* vol. 13 no. 3 (Winter 2011).

The afternoon session of 12 June and the morning session of 13 June featured the kind of wide panorama of studies of architecture and mathematics that have characterized the Nexus conferences since the very beginning. Moderated respectively by Kim Williams and Sylvie Duvernoy, presentations were made by Matthew Landrus, Vasco Zara, António Nunes Pereira, Eliseu Gonçalves, Teresa Marat-Mendes, Mafalda Teixeira de Sampayo and David Rodrigues, Jong-Jin Park, Marco Giorgio Bevilacqua, João Paulo Cabeleira, and Juergen Bokowski. These studies ranged from the historical to the contemporary, and dealt with themes as varied as urban design, design theory, architecture and music, proportional systems, optics and perspective, and morphology. This group of papers is scheduled for publication in the *Nexus Network Journal* vol. 13 no. 2 (Summer 2011).

The afternoon session of 13 June was dedicated to "Shape and Shape Grammars" and was moderated by Lionel March. Papers were presented by Sara Eloy and José Pinto Duarte, José Nuno Beirão, José Pinto Duarte and Rudi Stouffs, and by Mine Özkar. Unfortunately, George Stiny was prevented from presenting his paper in person by the untimely death of William (Bill) Mitchell, a well known and well liked authority on shape grammars. Prof. Stiny's paper was presented by Lionel March. This group of papers is scheduled for publication in the *Nexus Network Journal* vol. 13 no. 1 (Spring 2011).

The outstanding keynote presentations that concluded each day of the conference were given by Chris Williams, Lino Cabezas and Eduardo Souto de Moura. Chris Williams is a structural engineer who worked for Ove Arup and Partners prior to joining the Department of Structural Engineering at the University of Bath. His keynote talk was entitled "Patterns on a surface: the reconciliation of the circle and the square". Lino Cabezas teaches in the Departamento de Dibujo of the Facultad de Bellas Artes of the Universidad de Barcelona. His keynote talk was entitled "Ornament and structure in the representation of Renaissance architecture in Spain". Portuguese architect Eduardo Souto de Moura is originally from Porto, and worked in the offices of Álvaro Siza before opening his own practice. He is the designer of the Braga Municipal Football Stadium, among other internationally acclaimed works. His keynote talk discussed how materials convey certain kind of geometries in the construction of architectural space. The papers of Chris Williams and Lino Cabezas are scheduled for publication in the *Nexus Network Journal* vol. 13 no. 2 (Summer 2011).

The conference was enlivened by the many beautiful posters that lined the room where coffee breaks and lunches were served. Conference participants were free to chat with the creators of posters during the breaks.

Nexus conference participants who arrived the day before the conference began were offered the chance to attend the seminar entitled "Open Source" held at Porto's Casa da Música (designed by designed by the Dutch architect Rem Koolhaas with Office for Metropolitan Architecture and Arup-AFA), which featured engineer Cecil Balmond and others.

The conference concluded with a post-conference tour to seventh-century S. Frutuoso Chapel in Montélios and the Braga Municipal Stadium designed by Eduardo Souto de Moura. The post-conference tour was led by João Pedro Xavier.

The organizers of Nexus 2010 would like to take this opportunity to thank all those who worked so hard to make this conference a success, starting with the University of Porto, especially the Faculdade de Arquitectura da Universidade do Porto and the Centro de Matemática da Universidade do Porto. The conference was made possible by support from the Universidade do Porto, Fundação para a Ciência e Tecnologia and the Fundação Calouste Gulbenkian. We are grateful to the FAUP conference "dream team", led by Suzana Araújo, with the help of Pedro Matos, Nuno Machado, Conceição Noverça, Ludovina Vale, Anabela Menezes, João Valentim, Anselmo Laires, Joaquim Rocha, Rosa Ferreira, Ana Susete Maia and the students (who wore blue Nexus t-shirts) André Prata, António Martins, Guilherme Sepúlveda, Maria João Soares, and Vitório Leite. Eng. José Simões, from Serafim Pereira Simões e Successores, Lda. kindly provided the thank-you gifts we gave to the speakers, a nice pen holder designed by Eduardo Souto de Moura, and Mr. Francisco Mendes was responsible for the catering service for coffee breaks and lunches. We also wish to thank the members of the FAUP Executive Board and the ex-Dean of this Faculty, Prof. Francisco Barata, who encouraged this initiative from the beginning.

Lionel March

The Martin Centre
University of Cambridge
1-5 Scroope Terrace
Cambridge CB2 1PX UK
lmarch@ucla.edu

Keywords: shape grammar,
design theory

Letter from the Guest Editor

Forty Years of Shape and Shape Grammars, 1971 – 2011

Abstract. Guest Editor Lionel March introduces *Nexus Network Journal* volume 13 number 1 (Winter 2010) dedicated to Shape and Shape Grammar.

Introduction

To start with, a personal view and history. There were two of us at neighboring desks in the studio. Both of us had come up to Cambridge and were a year or so into reading the mathematical tripos when, independently, with the arrival of Leslie Martin as Professor of Architecture, we decided to move over to architecture. Christopher Alexander and I would occasionally walk along Kings Parade. Not surprisingly, having been born in Vienna, Chris would talk about Wittgenstein's *Blue and Brown Books* [1958]. I had read Wittgenstein's *Remarks on the Foundations of Mathematics* [1956] which had only recently been published from student notes taken from his lectures, including sharp interventions from Alan Turing. I had also come across Chomsky's early research work funded by the US Navy. This would have been in the late 1950s. Separately, in our subsequent careers, as an editor, I was privileged to promote rule-based 'shape grammars', and Alexander, with associates at the Center for Environmental Structure, University of California Berkeley, was to pursue and establish a 'pattern language'. Rules, grammar, and language are at the centre of Wittgenstein's philosophical questionings.

In the early 1970s, the Royal Institute of Architecture's Library Committee invited me to write a book relating the 'new' mathematics to architecture. In turn, I invited Philip Steadman to join me and *The Geometry of Environment* followed. As proofs were coming off the press, I came across a paper in the University's Computer Laboratory – George Stiny and James Gips' 'Shape Grammars and the Generative Specification of Painting and Sculpture' [1971] – just too late to reference. This paper started it all, and forty years later, the eight papers in this 2011 edition of the *Nexus Network Journal* provide a glimpse of progress today across the globe: from Stiny in the United States, to Li in Japan; from Özkar in Turkey, to Turkienicz in Brazil; and Duarte with colleagues in Portugal and the Netherlands.

Unfortunately, George Stiny was unable to present his paper in person at the 2011 conference because he was to deliver an address at the MIT memorial service for William J. Mitchell who had died just days before, on June 11. Readers of the *NNJ* will be aware of the considerable contribution to the promotion of 'mathematics and architecture' through computer-aided design that Bill Mitchell made during his career. Bill Mitchell had recommended George Stiny to me in a letter when I briefly held a Professorship in Systems Engineering at the University of Waterloo. On my subsequent appointment as Professor in Design at The Open University, I immediately offered George a lectureship. He accepted. It had been a blind appointment. We first met face to face on his arrival from California on a bitterly cold day in Woburn, England. His opening remark: "I agree with 95% of what you have written." Our conversation, since then, has never stopped.

Nexus Network Journal 13 (2011) 5–13   NEXUS NETWORK JOURNAL – VOL.13, No. 1, 2011 5
DOI 10.1007/s00004-011-0054-8; *published online* 22 February 2011

I used to quote Lord Kelvin: 'When you can measure what you are speaking about and express it in numbers, you know something about it, but when you cannot express it in numbers, your knowledge is of a meagre and unsatisfactory kind' (see [March 1976: 41]). But what numbers? Take a look at this example from Wittgenstein' *Remarks* ... :

> An addition of shapes together, so that some of the edges fuse, plays a very small part in our life. – As when

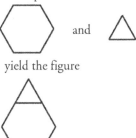

and

> yield the figure

> But if this were an *important* operation, our ordinary concept of arithmetical addition would perhaps be different [1956: VII-61].

How different? Typically, Wittgenstein asks a question, but does not follow through. For the past forty years, Stiny has sought answers to what he deemed to be no small part and a very important operation in design: how to compute with shapes. In this arithmetical expression: 3 plus 6, the sum equals 9; however 9 take away 3 gives back 6, or 9 take away 6 gives back 3. Yet, what happens with shapes? Taking away the triangle from the combined figure does not give back the hexagon. Or, take away the hexagon and the triangle has gone.

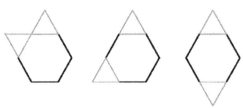

Clearly, the traditional numbers game is up. Traditionally, if apples on a plate are counted, each apple can be taken one at a time and transferred to another plate. That is, counting means subtracting one at a time. Let's go on. There are three ways of adding two triangles to the hexagon under symmetry – a new notion, absent in arithmetical addition.

Subtracting the triangles (in grey) gives us back the two we added, but where is the hexagon? Three new and distinct shapes occur in place of the hexagon. If we had subtracted the hexagon instead, no triangles would have been counted. With three triangles

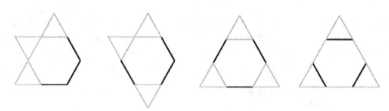

there are again three distinct figures, but with a difference. The second figure has a handed partner, or enantiomorph – yet another concept. The third figure has two counts of triangle. Three small triangles can be subtracted, or one large emergent, scaled-up triangle. Emergence, scaling – more concepts. Taken sequentially, only one of these answers can arise: the answer is ambivalent. Both three and one are correct in isolation. Four different shapes replace the hexagon. No triangles can be counted, if the hexagon is subtracted.

Four and five triangles added to the hexagon give similar ambivalent results. There is only one figure for five under symmetry, but two answers depending on whether small triangles are subtracted, or one large emergent triangle:

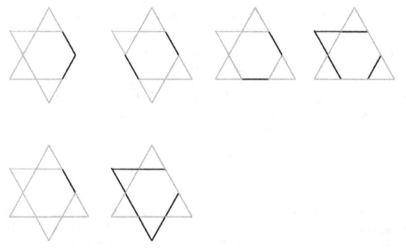

Things are more interesting with six triangles added. The combined figure has the symmetry of the hexagon. Six small triangles can be subtracted to leave no hexagon, or two large triangles can be subtracted from the hexagon which vanishes. Finally, the hexagon is subtracted leaving angles spanning the sides. How many triangles are there? 0, 2, 6, or – counting small and large triangles together – 8? The latter result can only be computed, if the deletion rule is applied in parallel.

This figure was looked at by Hermann Lotze: "We see it sometimes as two large triangles superimposed, sometimes as a hexagon with angles spanning its sides, sometimes as six small triangles stuck together at their corners" (cited in a footnote by William James [1950: 442-443]; also in [Stiny 2006: 152]). In brief, the arithmetic of shape is a matter of looking. There is no one way of seeing. There is ambivalence, ambiguity: the lifeblood of creativity. In related observations, I have used the relationship of two equilateral triangles to illustrate a point about Charles Babbage's (1791–1871) computing 'miracles' [March 1996].

Addition and subtraction of shapes are Boolean operations. Moving the original triangle to add to the hexagon exhibits Eulidean symmetry operations – translation, rotation, reflection, and scaling. Elementary topology is introduced with changes in dimension through a boundary operator. A 1-dimensional line is not made up of points, but it is bounded by its two 0-dimensional endpoints. A 2-dimensional face is bounded by 1-dimensional lines, a 3-dimensional solid by 2-dimensional faces. Take the boundary of the original figure

which comprises seven points. Take away the triangle – now defined by three vertices – and four points replace the hexagon. Or, take away the six vertices of the hexagon, and one point remains of the original triangle:

Note that these 0-dimensional shapes conform to normal arithmetic. Put another away, computation in ordinary arithmetic is inevitably 0-dimensional, even when graphic coordinates – (x, y) or (x, y, z) – boast 2- or 3-dimensions.

The inverse of the boundary operator, the co-boundary operator, fills in the 2-dimensional face of the figure

from which arbitrary numbers of equilateral triangles may be subtracted under scaling. The boundary of this face is the Boolean 'symmetric difference' of the triangle and hexagon, while the missing edge is their Boolean 'product'.

I run ahead of myself in what the English novelist Muriel Sparks calls 'narrative looping', or telling events in advance of their occurrence. To go back to the birth of shape grammars. James Gips' *Shape Grammars and their Uses: Artificial Perception, Shape Generation and Computer Aesthetics* was published by Birkhäuser in 1975; this was followed three volumes later in the same year by George Stiny's *Pictorial and Formal Aspects of Shape and Shape Grammars: On Computer Generation of Aesthetic Objects.* The former book derived from James Gips's doctoral thesis in Artificial Intelligence, Stanford University; the latter from George Stiny's doctoral dissertation submitted to UCLA's System Science Department. Both authors had been undergraduates at MIT where they attended classes given by Marvin Minsky and Naom Chomsky. It was in tackling an assignment designing handtools set by Minsky that the seeds of shape grammars were first sown. My own essay 'A boolean description of a class of built forms' was finally published in *The Architecture of Form* [March 1976]. The argument followed a personal interpretation of Max Newman's *Elements of the Topology of Plane Sets of Points* [1939]. I designed the cover of the first paperback edition of Newman's book for Cambridge University Press in 1964. Here is a fundamental difference. In my early work a line was understood to be a set of points, while for George Stiny, "all lines in a shape have finite, non-zero length" [1975: 4]. This harks back to Anaxogoras's (c. 500 BCE–428 BCE) principle of homeomerity and to its medieval reprise in the philosophical dispute between the atomism of Richard Grosseteste (1175–1253) and Thomas Bradwardine's (1290–1349) later view that a continuum is decomposable only into self-similar continua; that is, a line divides into lines, not points [March 2008].

The Cambridge philosopher of science, Mary Hesse [1963] had set out a position arguing that many theories in a particular field started as analogues of pre-existing theories in another field [Echenique 1972]. James Gips and George Stiny modeled their initial proposals on linguistic analogies. Explicitly: "Where a phrase structure grammar is defined over an alphabet of symbols and generates languages of strings of symbols, shape grammars are defined over an alphabet of shapes and generate languages of shapes" [Stiny 1975: 28]. Both authors define a shape grammar as a 4-tuple $\langle V_T, V_M, R, I \rangle$, where V_T is a finite set of terminal shapes, effectively a vocabulary; V_M is a finite set of markers, shapes distinguishable from those in V_T; R is a finite set of shape rules of the form $u \rightarrow v$ with appropriate provisos; I is called the initial shape. This formal apparatus is nowhere to be found in George Stiny's *Shape* [2006]. There is no requirement for a vocabulary, the marker set is reduced to a point, rules are classified into a variety of schema, and everything can start with the empty shape, a blank sheet of paper. The linguistic crutch has gone. Shapes are not symbols. Symbols operate under the identity relation, but shapes operate under the embedding – partial order – relation. Shape computations extend Turing's classical machine: 'recursion and identity' for symbols extend to 'recursion and embedding' for shapes. Symbols, like points, are 0-dimensional, while shapes comprising lines, faces and solids are 1-, 2-, and 3-dimensional respectively. A point is identified as its sole and only point, but a line is an uncountable myriad of embedded lines; faces and solids likewise. For a formal comparison of point set approaches and shapes, see Christopher Earl [1997].

In 1978, George Stiny and James Gips published *Algorithmic Aesthetics: Computer Models for Criticism and Design in the Arts.* A theme issue on design and language of *Environment and Planning B: Planning and Design* was published in 1981. This issue included Stiny and March's 'Design machines' [1981], Downing and Flemming's 'The bungalows of Buffalo' [1981], Koning and Eizenberg's 'The language of the prairie: Frank Lloyd Wright's prairie houses' [1981]. In 1994, Terry Knight published *Transformations in Design: A Formal Approach to Stylistic Change and Innovation in the Visual Arts.* This book was the first to present shape grammars in a form suitable for pedagogical purposes as well as presenting the author's original research into stylistic change, for example, in the meander patterns of Greek pottery; the paintings of George Vantongerloo and of Fritz Glarner; and the transformation of the prairie house to the architecture of the Usonian homes by Frank Lloyd Wright. As an introductory text, Terry Knight's remains unsurpassed. In 1999, a special design and computation edition of *Environment and Planning B: Planning and Design* was published, guest edited by myself and Terry Knight. Apart from papers by myself, Knight and Stiny, the issue contained Djordje Krstic's setting of shape grammars into the frame of multi-sorted algebras; Scott Chase's demonstration of the use of shape grammars to permit 'emergence' in geographical information systems; Mark Tapia's computer implementation of a simple shape grammar; and Athanassios Economou's examination of the symmetries of Froebel kindergarten blocks employing Polya's counting theorem (see 'Kindergarten grammars: designing with Froebel's building gifts' [Stiny 1980]).

At the turn of the century, shape grammars were prominently featured in Erik K. Antonsson and Jonathon Cagan's *Formal Engineering Design Synthesis* [2001]. George Stiny's *Shape: Talking about Seeing and Doing* [2006] revisited familiar ground and radically revised the very foundations of computing with shapes. Two years later, guest edited by me [March 2008], *Planning and Design* reprinted seven selected shape grammar papers starting with George Stiny's 'Two exercises in formal composition' [1976] – the very first shape grammar paper that I published, as founding editor, in the third volume of *Environment and Planning B*. Of possible and particular interest to *NNJ* readers, will be Stiny and Mitchell's 'The Palladian grammar' [1978]. Papers by Flemming [1987], Knight [1989], and Earl [1997] are also included in the selection; together with a letter from Stiny [1982] in which he makes the important distinction between set grammars and shape grammars. Given the location of the conference recorded here – Porto, Portugal – it seems appropriate that this special issue of *Planning and Design* concluded with José Duarte's 'Towards the mass customization of housing: the grammar of Siza's housing at Malagueira' [2005], based on his MIT doctoral dissertation. This volume also included a perceptive review of *Shape* by Earl, reprinted the following year [Earl 2009]. For reasons of space, the volume regrettably did not include a reprint by Krishnamurti, who had contributed over fifteen noteworthy papers to *Planning and Design* since 1978, many related to the computer implementation of shape grammars.

To conclude: this present issue of the *NNJ* contains eight papers that resulted from the 2010 Nexus conference section on Shape and Shape Grammars. George Stiny's paper provides the keynote for the section. Among the twelve contributors, José Duarte, Mine Özkar, and Andrew Li are former doctoral students of Stiny's at MIT; while Rudi Stouffs was a doctoral student under Ramesh Krishnamurti's supervision who, himself, had been Stiny's postdoc at The Open University. José Duarte is involved in four of the papers and his achievement and enthusiasm in bringing shape grammars to Portugal deserves recognition.

In selecting the papers, it seemed appropriate to spread themes as widely and representatively as possible. George Stiny provides further theoretical insights developing the concept of *schemas* first mooted in 'How to calculate with shapes' [2001] and advanced in *Shape* [2006], with some deft hints at Alberti's prescience in *Elementi di pittura* [Williams, March, Wassell 2010]; Mine Özkar, alert to visual schemas, having produced more than a thousand illustrations for Stiny's *Shape,* demonstrates examples of pedagogical applications in the architectural studio – a matter Stiny stresses on closing his keynote; Andrew Li takes up the theme of shape language and style in classical Chinese architecture. José Duarte collaborates with Filipe Coutinho and Mário Krüger – editor and commentator on the current Portuguese translation of Alberti's *De re aedificatoria* – in using shape grammars and descriptions to 'decode Alberti 'and to employ the results in arguments concerning influences on architecture during the Portuguese Renaissance. The paper by Alexandra Paio and Benamy Turkienicz uses a shape grammar to visualize the written rules of composition of colonial Portuguese cities and their subsequent cartographic presentation.

José Duarte, in the first of his three remaining collaborative pieces, joins with Sara Eloy in using shape grammar formalism as an aid to the rehabilitation of housing stock among *rabo-de-bacalhau* in Lisbon. In part, they incorporate graph techniques from 'space syntax'. The paper contributes to the concern for conservation and rehabilitation first broached in shape grammar literature by Flemming [1987]. At the contemporary urban scale, Duarte collaborates with José Beirão and Rudi Stouffs from TU Delft in creating a design system involving a parametric shape grammar with descriptions to generate urban block layouts within a defined spatial region. In their discussion, they evoke Alexander's pattern language as manifest in software design. José Duarte, finally, works with Maria da Piedade Ferriera and Duarte Cabral de Mello in presenting a novel example of a kinetic shape grammar simulating human body movements. The paper presents rules within a confined set of yoga movements for demonstration purposes; but it also presents a fascinating historical survey of the study of human movement.

References

ANTONSSON, Erik K. and JONATHON Cagan. 2001. *Formal Engineering Design Synthesis.* Cambridge: Cambridge University Press.

CHASE, Scott. 1999. Supporting emergence in geographic information systems. *Environment and Planning B: Planning and Design* **26**: 33-44.

DOWNING, F. and Ulrich FLEMMING. 1981 The bungalows of Buffalo *Environment and Planning B: Planning and Design* **8**: 269-294.

DUARTÉ, José. 2005. Towards the mass customization of housing: the grammar of Siza's housing at Malagueira. *Environment and Planning B: Planning and Design*. 32: 347-380.

EARL, Christopher. 2009. Review of *Shape: Talking about Seeing and Doing. Environment and Planning B: Planning and Design* **36**: 186 ff.

―――. 1997. Shape boundaries. *Environment and Planning B: Planning and Design* **24**: 669-687.

ECHENIQUE, Marcial. 1972. Models: a discussion. Pp. 164-174 in *The Architecture of Form*, Lionel March, ed. Cambridge: Cambridge University Press, 1976.

ECONOMOU, Athanassios. 1999. The symmetry lessons from Froebel building gifts. *Environment and Planning B: Planning and Design* **26**: 75-90.

FLEMMING, Ulrich. 1987. More than the sum of parts: the grammar of Queen Anne houses. *Environment and Planning B: Planning and Design* **14**: 323-350.

GIPS, James. 1975. *Shape Grammars and their Uses: Artificial Perception, Shape Generation and Computer Aesthetics* (Interdisciplinary Systems Research series 10). Basel and Stuttgart: Birkhäuser Verlag.

HESSE, Mary. 1963. *Models and Analogues in Science.* London: Sheed and Ward.

JAMES, William. 1950. *The Principles of Psychology: Volume 1* (1890). New York: Dover.

KNIGHT, Terry. 1999. Shape grammars: six types. *Environment and Planning B: Planning and Design* **26**: 15-32.

―――. 1994. *Transformations in Design: A Formal Approach to Stylistic Change and Innovation in the Visual Arts.* Cambridge: Cambridge University Press.

―――.1989. Color grammars: designing with lines and colors. *Environment and Planning B: Planning and Design* **16**: 417-449.

KONING, H. and J. EIZENBERG. 1981. The language of the prairie: Frank Lloyd Wright's prairie houses. *Environment and Planning B: Planning and Design* **8**: 295-324.

KRISTIC, Djorde. 1999. Constructing algebras of design. *Environment and Planning B: Planning and Design* **26**: 45-58.

MARCH, Lionel. 1976. *Architecture of Form*. Cambridge Urban and Architectural Studies (No. 4). Cambridge: Cambridge University Press.

———. 2008. Guest editorial: The shape grammarist's voice is ineluctably personal. *Environment and Planning B: Planning and Design. Shape grammars—a SIGGRAPH 2008 conference issue of selected reprints*: 1-5.

———. 1999. Architectonics of proportion: a shape grammatical depiction of classical theory. *Environment and Planning B: Planning and Design* **26**: 91-100.

———. 1996. Babbage's miraculous computation revisited. *Environment and Planning B: Planning and Design* **23**: 369-376.

———. 1976. A boolean description of a class of built forms. Pp. 41-73 in *The Architecture of Form*. Cambridge: Cambridge University Press, 1976.

MARCH, Lionel and Philip STEADMAN. 1971. *The Geometry of Environment*. London: RIBA Publications.

NEWMAN, M. H. A. 1939. *Elements of the Topology of Plane Sets of Points*. Cambridge: Cambridge University Press.

STINY, George. 2006. *Shape: Talking about Seeing and Doing*. Cambridge MA: MIT Press.

———. 1991. The algebras of design. *Research in Engineering Design* 2: 171–181.

———. 1982. Spatial relations and grammars. *Environment and Planning B: Planning and Design* **9**: 113-114.

———. 1980. Kindergarten grammars: designing with Froebel's building gifts. *Environment and Planning B: Planning and Design* **7**: 409-462.

———. 1976. Two exercises in formal composition. *Environment and Planning B: Planning and Design* **3**: 187-210.

———. 1975. *Pictorial and Formal Aspects of Shape and Shape Grammars: On Computer Generation of Aesthetic Objects* (Interdisciplinary Systems Research series 13). Basel and Stuttgart: Birkhäuser.

STINY, George, and James GIPS. 1971. Shape grammars and the generative specification of painting and sculpture. *Proceedings of IFIP Congress '71*. Amsterdam: North Holland Publishing Co. Also in *The Best Computer Papers of 1971*, O. R. Petrocelli, ed. Philadelphia: Auerbach, 1971.

———. 1978. *Algorithmic Aesthetics: Computer Models for Criticism and Design in the Arts*. Berkeley and Los Angeles: University of California Press.

STINY, George and Lionel MARCH. 1981. Design machines. *Environment and Planning B: Planning and Design* **8**: 245-256.

STINY, George and William MITCHELL. 1978. The Palladian grammar. *Environment and Planning B: Planning and Design* **5**: 5-18.

TAPIA, Mark. 1999. A visual implementation of a shape grammar. *Environment and Planning B: Planning and Design* **26**: 59-74.

WILLIAMS, Kim, Lionel MARCH and Stephen R. WASSELL, editors. 2010. *The Mathematical Works of Leon Battista Alberti*. Basel: Birkhäuser.

WITTGENSTEIN, Ludwig. 1956. *Remarks on the Foundations of Mathematics*. Oxford: Basil Blackwell.

———. 1958. *Blue and Brown Books*. Oxford: Basil Blackwell.

About the author

Lionel March holds a BA (Hons) in Mathematics and Architecture, Diploma in Architecture, MA, and Doctor of Science (ScD) from the University of Cambridge. He has held fellowships with the Royal Society of Arts, the Institute of Mathematics and its Applications, and the Royal College of Art. He was founding director of the Centre for Land Use and Built Form Studies, now the Martin Centre for Architectural and Urban Studies, University of Cambridge. He has held

Professorships in Engineering (Waterloo, Ontario), Technology (The Open University), Architecture and Urban Design, and Design and Computation (University of California, Los Angeles). He was Rector and Vice-Provost (Royal College of Art, London). He was co-editor with Sir Leslie Martin of the twelve volume Cambridge Architectural and Urban Studies, Cambridge University Press. He was founding editor in 1967 of the bi-monthly refereed journal *Environment and Planning B*, now *Planning and Design*.

Relevant recent publications include *Architectonics of Humanism: Essay on Number in Architecture* (1999); 'Renaissance mathematics and architectural proportion in Alberti's *De Re Aedificatoria*', *Architectural Research Quarterly* (*ARQ*), 2/1, 1996, 54-65; 'Proportional design in L. B. Alberti's Tempio Malatestiano, Rimini', *ARQ*, 3/3, 1999, 259-269; 'Palladio, Pythagoreanism and Renaissance Mathematics', *Nexus Network Journal*, 10/2, 2008, 227-243. Lionel March is currently Visiting Scholar, Martin Centre for Architectural and Urban Studies, Department of Architecture, University of Cambridge, and emeritus member, Center for Medieval and Renaissance Studies, University of California, Los Angeles.

George Stiny

Department of Architecture, 10-419
Massachusetts Institute of Technology
77 Massachusetts Avenue Cambridge,
MA 02139 USA
stiny@mit.edu

Keywords: shape grammar, design
theory, rules, schemas, embedding,
recursion, hierarchy

Research

What Rule(s) Should I Use?

Presented at Nexus 2010: Relationships Between
Architecture and Mathematics, Porto, 13-15 June
2010.

Abstract. Rules in schemas apply to shapes in terms of
embedding. This melds calculating with art and design,
and has many uses in teaching and in practice.

Introduction

Every time I talk about shape grammars, and art and design, I'm asked exactly the
same question – usually more than once:

Q1 What rule(s) should I use?

And Q1 is easy to answer:

A1 Use any rule(s) you want, whenever you want to.

There are plenty of rules to choose from and, because rules depend on embedding, any
rule you use will do the job, or let you go on to something new. Calculating with shapes
is an open-ended process – like art and design. You're always free to try another rule.

This is the best answer I know. It extends calculating in Turing's way with symbols
using recursion and identity – in other words, calculating in the canonical way with all-
or-none bits (0's and 1's) – to calculating in my way with shapes using recursion and
embedding. Turing's kind of calculating is logical and combinatory – it's just counting.
But it can get hung up and crash because there may be no symbols or way of combining
them to go on. My kind of calculating is phenomenal and involves seeing. Then you can
always go on because whatever you see – anything at all – is OK. There are no non
sequiturs or blind alleys. This doesn't matter. My answer is rarely greeted enthusiastically.
Maybe my interlocutors want to hear something else. There are many things to ask.
Their real question may go more like this:

Q2 Can you tell me how to design an x, or what to do next when I'm
designing an x?

The problem with Q2 is that it's usually too hard – at least for me. I probably don't
know how to design an x. I may know something about shape grammars and calculating
with shapes and rules, but this doesn't mean I know the rules to use for any x, y, or z.
What I suggest may not be anything you would do or anyone else would try. I may be
wrong. The only promise is that calculating goes on whatever rules you try – right or
wrong for x, y, or z, rules always work. Perhaps what's wanted is a little magic, so that
designing isn't necessary. This may be too much to ask. I'm not a sorcerer, even if shapes
and rules may sometimes seem preternatural. But there's a practical question in between
these two extremes that may repay discussion:

Q3 How can I cook up rules to match what's in art and design?

My answer to Q3 subsumes other things I've tried in the past, for example, my
scheme for kindergarten grammars using spatial relations defined in a given vocabulary of

shapes. This is still a valuable technique in teaching and practice, but now I want something more freewheeling. Instead of a vocabulary of shapes and spatial relations, I'll start with a small lattice of schemas and combine them to get other schemas as they're needed. The rules in these schemas incorporate key ideas in algebras of shapes, from various branches of mathematics. These include embedding and parts (Boolean algebra), transformations (geometry), and shape boundaries (topology). This lets me calculate with shapes, so that what I see makes a difference in what I do. This is reason enough for my schemas, but there's another reason, too, that may make them inevitable. The rules in my schemas are easy to find throughout art and design. Pragmatic evidence matters a lot when deciding what rules to try. Rules that work, work.

Getting started

Different algebras of shapes are enumerated in this familiar array:

$$U_{00} \quad U_{01} \quad U_{02} \quad U_{03}$$
$$U_{11} \quad U_{12} \quad U_{13}$$
$$U_{22} \quad U_{23}$$
$$U_{33}$$

In any algebra U_{ij}, the index i gives the dimension of the basic elements that make up shapes:

dimension i =	basic elements
0	points
1	lines
2	planes
3	solids

All basic elements, except for points, have boundaries that are shapes. The basic elements in these shapes have smaller dimension:

basic elements	boundary shapes made up of
solids	planes
planes	lines
lines	points
points	none

The index j is greater than or equal to the index i. The index j gives the dimension of the space in which basic elements are defined, and combined and moved around, or, equivalently, the dimension of the transformations that apply to shapes. Shapes may be added or subtracted, and changed by the transformations. Shapes are parts of shapes that can be divided anywhere, and shapes fuse without divisions – once shapes go together they disappear in the result.

There's a lot more than this, but this is plenty for now. With seeing as a guide, you can't go wrong with shape grammars. You don't have to define everything formally when you can follow your eyes. In fact, formal definitions have to live up to seeing in every way to be of any value at all. Your eyes are always the test. So let's take a look at my third question and not worry too much about being perfectly rigorous:

Q3 How can I cook up rules to match what's in art and design?

What schemas are in my lattice and what rules do they define? I have three sorts of operators in mind that let me define simple rules to move within algebras of shapes and between algebras of shapes. I can use these rules to go from one shape to another shape of the same kind, and to link shapes of different kinds. My operators involve parts of shapes, transformations, and boundaries:

(1) prt(x) determines a part of a shape x,

(2) t(x) transforms x into another shape, and

(3) b(x) gives the boundary shape of x.

The rules I want are in the schemas in this lattice:

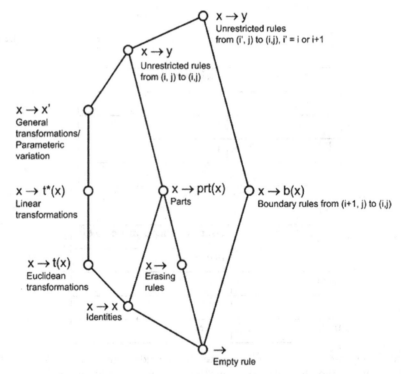

The lattice is defined in the customary way for subsets. Two schemas are connected if the rules in the lower one are also rules in the higher one. Rules for parts, transformations, and boundaries provide some of the rudiments to calculate with shapes in art and design.

Let's start at the bottom of my lattice and work up to the top, from restricted schemas to more general ones. I'll try to describe rules in terms of the least inclusive schemas in which they're defined. The idea is to show how little it takes to do a lot – in accord with Ockham's razor. The empty rule

$$\rightarrow$$

is just the arrow that glues shapes together in spatial relations to define rules. The empty rule is as spare as you can get. It doesn't look like much because its left side and its right side are the empty shape. This captures the idea of an action – an unresolved impulse – without saying what to see or what to do. The empty rule closes off my lattice, and in this sense, it's merely a technical device. But the empty rule may be useful in some additional ways, as well – for example, to define rules that apply in parallel to calculate in composite algebras that are the products of others. This allows for multiple shapes and sundry stuff to interact in various ways that may change as calculating goes on. The relations that make sense now may be nonsense later.

The schema

$$x \rightarrow x$$

and the schema

$$x \rightarrow$$

are both directly above the empty rule. The one contains identities to fix the biggest parts of shapes, and the other, erasing rules for the smallest parts of shapes. Their top

$$x \rightarrow prt(x)$$

contains rules that pick out parts of shapes whatever their size – biggest, smallest, or in between. Together, rules of these three kinds show that shapes can be divided into parts in any way you like. Parts can be traced or cut out from here, there, or anywhere in any number of pieces. These rules are particularly useful when it comes to decomposing shapes into meaningful structures of various kinds, for example, into hierarchies or finite topologies (Boolean algebras and Heyting algebras). In shape grammars, these structures are defined on the fly as a result of what rules do, and they can alter erratically as rules are tried. Hierarchies and the like are retrospective, and aren't a prerequisite for calculating. It's just the reverse. The parts that are resolved in shapes are the outcome of calculating, and how these parts are related is determined by calculating, too. In shape grammars, no vocabulary of parts or structure of any kind is a prerequisite.

Identities in the schema $x \rightarrow x$ may be of special note. Many find them useless, because they apply to shapes to produce exactly the same ones. It seems that there's no reason to use identities to calculate. What good is calculating without effect – one shape C repeats monotonously with no relief:

$$C \Rightarrow C \Rightarrow ... \Rightarrow C$$

But this kind of conspicuous redundancy isn't really vacuous. In many ways, identities are at the root of shape grammars and their use in art and design. Identities let me change things without lifting a finger, simply looking again to find new parts and define new relations. And this may be pretty dramatic, as, for example, in R. Mutt's (or is it Marcel Duchamp's?) infamous *Fountain*. Mr. Mutt opens our eyes with identities, and Duchamp and his friends – Beatrice Wood and Henri-Pierre Roché – say so in the ultimate issue of *The Blind Man*. (The blind man, indeed, not seeing in one way but seeing in another.)

The identities in the schema x → x also show how transformations work. An identity is a rule in the schema

$$x \rightarrow t(x)$$

if t is the identity transformation (t(x) = x), but t may be any other Euclidean transformation, as well. The transformation rules in the schema x → t(x) apply recursively to t(x), and in surprising ways when there's embedding. It's magic! For example, I can use transformation rules to rotate or reflect shapes by translating their parts. This isn't the kind of geometry I learned in school. Or I can turn pinwheels to rotate points that are fixed as rules are tried. That is to say, I can move points without any obvious way of moving them. It's a neat trick that's fun to do over and over again in various ways.

Of course, the Euclidean transformations aren't the end of it. They're included with affine transformations in the linear (projective) transformations in the schema

$$x \rightarrow t^*(x)$$

Linear transformations t* provide the foundation for perspective in drawing and painting. They're key in architectural practice and have many important uses elsewhere in design.

General transformations are given by x′ in the schema

$$x \rightarrow x'$$

The transformation rules in the schema x → x′ allow for parametric variation of any sort – linear and not. They let me explore designs with shape schemas like Goethe's *Urpflanze*, too – but perhaps more generally, because organs (unit parts: atoms, bits, building blocks, cells, components, constituents, etc.) aren't predefined to permute and transform. The rules in my schemas use embedding to redefine parts again and again on the fly – in fact, every time a rule is tried. That's how rules match seeing. And that's why they work so well in art and design, and why they do far more than anyone expects.

The schema

$$x \rightarrow b(x)$$

is also above the empty rule → . The boundary rules in this schema let me go from shapes with basic elements of dimension i+1 to shapes with basic elements of dimension i, so that I can use all kinds of shapes at the same time when I calculate. These rules mark the ends of lines, or outline shapes made up of planes, as an artist or designer might do this in a drawing.

Unrestricted rules finish off my lattice. They're in the twin schemas

$$x \rightarrow y$$

Like the empty rule → at the bottom, they're needed for technical reasons. But they have explanatory content, as well. The lower schema only contains rules with shapes x and y in a single algebra with basic elements of dimension i. The top schema also contains rules with shapes x and y in two algebras with basic elements of dimensions i+1 and i. This is why rules for parts and transformations, and rules for boundaries are different. A little later, I'll show how to define unrestricted rules using the schema for parts, x → prt(x).

Going on

The schemas in my lattice are just the beginning of the story, although I sometimes think that they're more than enough to keep me going for a long, long time. They illustrate the novel properties of shape grammars in some really easy and surprising ways. There's more art in them than anyone will ever see. The schema x → x for identities may be all I need to prove this. Mr. Mutt would probably agree. Nonetheless, there are additional kinds of schemas to define rules that are hard to avoid in art and design. They're very intuitive rules that seem almost inevitable in teaching and practice – rules for the kinds of things almost everyone tries to do.

One way to define additional schemas is to take the inverses of the schemas in my lattice. Flipping the sides of schemas forms their inverses. Thus for the schema (or rule) x → y, the inverse is y → x. Inverses for some of the schemas in my lattice are given in this table:

schema	x→	x→prt(x)	x→t(x)	x→b(x)
inverse	→x	prt(x) →x	t(x) →x	b(x) →x
alternate form		x→prt^{-1}(x)	x→t^{-1}(x)	x→b^{-1}(x)

All of these inverses are pretty interesting, with precedents in art and design. Only the one for transformations, t(x) → x, may be superfluous, because the inverses of transformations are transformations themselves. Any rule in the inverse t(x) → x is already a rule in the schema x → t(x) in my lattice. This is explicit in the alternate form x → t^{-1}(x) – but more on the use of alternative forms one paragraph down.

The three schemas → x, prt(x) → x, and b(x) → x are especially nice. The rules in the schema → x let me act without seeing anything at all, whether something is there or not. I can splash a wall with colors and go on from what I see in the sundry blots and stains, in the way Leonardo da Vinci urges – "if you consider them [blots and stains] well, you will find really marvelous ideas." I may try a rule in → x and identities in the schema x → x to get what I want, or I can use → x more than once with x → x in a coordinated way. (This looks like action painting, where → x may be a procedural description for shapes – splash like so to get an x.) Perhaps there are blots and stains that need to be completed to get the right result – a contour needs to be closed up or an area needs to be filled in. Then rules in the schema prt(x) → x let me find parts and add what's missing or left out. Or maybe there's a contour that should be filled in with a boundary rule in the schema b(x) → x – something like drawing in a coloring book to shade in an open area and erase its outline, as well.

The inverses in my table are useful schemas, but they're awkward for what I want to try next in a mathematical fashion. For this, I need another way of saying the same thing. These alternate forms are already in place. Instead of flipping schemas, I'll use an inverse notation for parts and boundaries that copies my notation for transformations. Instead of the schema

$$\text{prt}(x) \rightarrow x$$

that reads "part of x goes to x," I'll use the alternate schema

$$x \rightarrow \text{prt}^{-1}(x)$$

where "x goes to a shape with x as a part." The variable x has twin values in these schemas – the rule A → B is in prt(x) → x for x = B, and in x → prt⁻¹(x) for x = A. (The duality for parts is clear in algebra, topology, etc. Then prt(x) corresponds to an ideal, and prt⁻¹(x) to a filter.) Equally, the schema

$$b(x) \rightarrow x$$

that reads "the boundary of x goes to x," has the alternate schema

$$x \rightarrow b^{-1}(x)$$

in which "x goes to a shape with the boundary x." This is something like using passive verbs in place of active ones. It's usually safe to go from one voice to another with no serious change in meaning, even if there always seems to be a mellifluous choice – I guess pleasing sounds also sound right.

I can combine the schemas in my lattice and their inverses to define compound rules that are widely used in art and design. I'll try two methods of combination: composition and addition. This isn't the end of it but only a start that may need to be augmented or revised, as experience demands. I've just started to play with this stuff. There's more to explore to be fully confident of the results. But this is no reason not to see how far I can go with what I've got to get others to wander around in this region to see what kinds of schemas and rules they can find.

Suppose I outline a shape x, as I look at prt(x). (There are many ways to think about this. Terry Knight tries "representational ambiguity" – my next rule includes a rule she defines.) Maybe I see a random patch

in the clouds or in smudges on a dirty window, and I draw the square

on a sheet of paper on my desk. Then I've used the rule

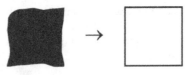

in the schema

$$prt(x) \to b(t(x))$$

The transformation t takes

from its part in the clouds to my desk – no doubt, with a change in scale – where the boundary operator b gives me the outline I want

(More explanation may be superfluous, but detail is welcome. My rule in the schema $prt(x) \to b(t(x))$ erases the original image – the wind blows it away, or it rains, which works for clouds and smudges alike. To leave the image alone, I need a rule in the schema $prt(x) \to prt(x) + b(t(x))$. This new schema is defined with addition. Then there are rules in the schema $x \to$ to erase clouds or smudges to keep only my drawing. But the schema $prt(x) \to b(t(x))$ is the crux of what I want to show now. Note, too, that color, shading, texture, etc. in clouds and smudges are something to calculate with, in schemas and rules. This takes weights. Their use is an open area of untapped promise.)

The schema $prt(x) \to b(t(x))$ is neither in my lattice nor the inverse of a schema that is, but it's easy to define by composition using such schemas:

$$
\begin{array}{l}
x \Rightarrow prt^{-1}(x) \Rightarrow t(prt^{-1}(x)) \Rightarrow b(t(prt^{-1}(x))) \\
\underline{} \\
x \to prt^{-1}(x) \\
x = x \\
\qquad \underline{} \\
\qquad x \to t(x) \\
\qquad x = prt^{-1}(x) \\
\qquad\qquad \underline{} \\
\qquad\qquad x \to b(x) \\
\qquad\qquad x = t(prt^{-1}(x)) \\
\underline{} \\
x \to t(prt^{-1}(x)) \\
\underline{} \\
x \to b(t(prt^{-1}(x)))
\end{array}
$$

The idea is to extend a series of expressions starting with x, to get a new schema. The schemas I already have are applied in this process, so that their left sides correspond to expressions of increasing size. A schema is defined if I put an arrow between x and any expression in the series. (Calculating with schemas isn't the only way to get new ones. I can combine expressions that I've built up by composing my operators and their inverses.

This gives the schema prt(x) → b(t(x)). But it may pay to calculate first, to describe schemas in a uniform way that helps to classify them, and to see and say what they do.)

The schema

$$prt(x) \rightarrow b(t(x))$$

in the form

$$x \rightarrow b(t(prt^{-1}(x)))$$

is a kind of speedup schema that allows for multiple actions at once. The rule

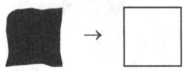

in this schema may look simple, but there's a lot to it – at least in terms of the schemas in my lattice. But this may seem odd. It may be more intuitive to define the rule directly instead of defining three other rules to get this easy result. Should parsimony and formal scruples limit the schemas I use? No, I don't think so. I'm not keen on any kind of morality when it comes to calculating, or to art and design. Schemas should define rules according to what artists and designers see and do, without a lot of work. I'm a big fan of exuberant (prodigal) redundancy, especially when there are easy results. But redundancy may make it hard to see what's going on and to talk about how various schemas and rules fit together. Defining some things from others may repay the effort. It organizes schemas and rules, and it helps to remember ones that have been useful in the past and to suggest new things to try. Thinking up schemas from scratch may be the hard way to get what you want, when you can change schemas and combine them to take advantage of what you know – experience counts. I'm a big fan of composition, as well.

A lot of redundancy is built into composition from the start. Suppose I work backwards from the square

on the piece of paper on my desk to the shape

in the clouds. Then I get the schema

$$x \rightarrow prt(t(b^{-1}(x)))$$

and its inverse

$$prt(t(b^{-1}(x))) \rightarrow x$$

defines the rule

Composition is pretty flexible – there may be many ways to do the same thing with a different twist. Maybe you can find another schema to define my rule. But being clever isn't the goal, seeing is. Embedding is still the key. Without embedding, schemas – no matter how clever – don't go very far.

All sorts of schemas are defined using composition. The schema

$$x \rightarrow b(b(b(x)))$$

lets me go from a solid to its vertices, and the schema

$$x \rightarrow b(b(x))$$

from a solid to its edges, or from a plane to its vertices. And inverses run through Kandinsky's point and line to plane in many ways. The schema

$$x \rightarrow b^{-1}(prt^{-1}(x))$$

turns points into lines, lines into planes, etc. – say, a square plane into a solid cube. But compositions may not be novel. The schema $x \rightarrow prt(x)$ includes the composite schema $x \rightarrow prt(prt(x))$. A part of a part of a shape x is a part of x. In the same way, the schema $x \rightarrow t(t(x))$ is a frill – transformations are closed under composition.

The schema $x \rightarrow prt(x)$ and its inverse have a neat composition

$$x \rightarrow prt^{-1}(prt(x))$$

in a universal schema for all rules – let $prt(x)$ be the empty shape that's a part of every shape. This shows how to get everything from nothing, but it says precious little about specific rules. The schemas in my lattice, and inverses, composition, and addition aren't just math. The aim is to hone schemas to describe and classify rules in useful (heuristic) detail – at least for the time being – for teaching and practice in art and design. Rules are all the same in my universal schema – there's scant to tell them apart in terms of the shapes they contain, or to describe the relations they may define.

Composition is hugely productive. Nonetheless, schemas combine in other ways, as well. I can use the plus sign (+) to link them in additions, so that the resulting schemas are explicitly recursive. For example, I may want to design a symmetrical pattern. Then rules in the schema

$$x \rightarrow x + t(x)$$

work perfectly, when the transformations t are the generators of a symmetry group. The identity $x \rightarrow x$ and the schema $x \rightarrow t(x)$ do the trick in the sum

$$
\begin{array}{l}
x \rightarrow x \\
\underline{+\ \ x \rightarrow t(x)} \\
\quad x \rightarrow x + t(x)
\end{array}
$$

The rules in this addition schema are in the inverse x → prt⁻¹(x), and they apply recursively to x and t(x). I used them nearly 35 years ago in "Two Exercises in Formal Composition." But my exercises weren't compositional in the logical sense in which parts are independent in combination – this takes labels or like devices to skirt ambiguity. The idea was to give rules for synthesis/analysis, so that this is a continuous process in design. Addition rules are further described in my scheme for kindergarten grammars, with a host of examples from Froebel's building gifts and tablets, H. G. Wells's imperial *Floor Games*, and basic design curricula of various kinds. And inverses of addition rules – subtraction rules – are defined and tried in many ways, too. Knight calls the rules in the schema x → x + t(x) "basic rules" and enumerates their properties when embedding is limited to identity. She uses basic rules – alone and also in "color grammars" – in her studio on computational design at MIT.

(Deciding what to say about schemas is never obvious – schemas can be described in different ways for different purposes that may or may not be meaningfully related. For example, both x → t(x) and x → x + t(x) are special versions of the addition schema x → prt(x) + t(x), when prt(x) is the empty shape or x. This presents new possibilities to explore, some of everyday use in design. Suppose I want to move the left-most square in the shape

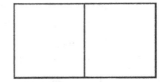

and leave the right square intact. But what happens if I try to translate the rectangle in the shape and keep the two squares, or if I alter one square and keep the other one and the rectangle? How many squares and rectangles are there? Are you sure? Do weights make a difference? How about composite algebras? It's no use trying to cover everything all at once. In what follows, I consider x → t(x) and x → x + t(x) in terms of transformations only. This is far from the end of it – there are parts and other things to account for, too.)

There are new rules in the more general schema

$$x \to x + x'$$

to handle movement, and parametric or alternative views in sums. This is clear in art and design – for example, in traditional Chinese window grilles with their "checkerboard" lattice designs (Euclidean transformations may do in this case), cubist paintings, and Aalto, Scharoun, Klee, and Duchamp (*Nude Descending a Staircase* (No.2)). And then the fully general schema

$$x \to x' + x''$$

is good for Chinese ice-ray lattices. The rules for ice-rays divide polygons into polygons – triangles, quadrilaterals, pentagons, etc. into different pairs of these figures. A polygon x goes to a sum of its transformations x' and x'' in a fractal-like fashion

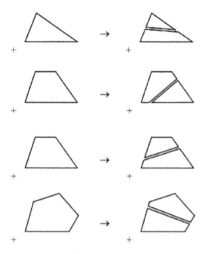

This may be better expressed with division rules in the schema

$$x \rightarrow \text{div}(x)$$

The schemas $x \rightarrow x' + x''$ and $x \rightarrow \text{div}(x)$ produce the same result when the operator $\text{div}(x)$ divides x into x' and x''. Addition and division look alike, although some – notably Christopher Alexander – try to make a big deal out of this "timeless" distinction. (Timelessness is no guarantee that a distinction isn't vacuous. In fact, it seems that many time-honored distinctions fail when examined in terms of shapes and rules. This is another way of showing how embedding works and what it does, and why seeing instead of counting is a good way to calculate.) Also notice that $x \rightarrow \text{div}(x)$ shows how to construct ice-ray windows. It's easy to imagine an artisan following the rules it defines to divide polygons with sticks, adding them in this way or that between the sides of polygons to keep everything rigid, from the empty window frame to the penultimate open areas. No one doubts that calculating with rules is rational. For ice-rays, it seems, this is rational design, too. Maybe I should include $x \rightarrow \text{div}(x)$ in my lattice. It's obviously useful, and it has widespread intuitive appeal. It's no trouble to add in. I guess I'd put it above the schema $x \rightarrow x$, so that identities are division rules, and in between one of the three schemas for transformation rules and the schema for parts, $x \rightarrow \text{prt}(x)$. The only downside is an increase in redundancy with respect to addition and the schema $x \rightarrow x' + x''$, and perhaps, too, the loss of some elegance in a sloppier lattice that includes more than algebras of shapes require.

The division schema $x \rightarrow \text{div}(x)$ always surprises me: its results are everywhere I look. Rules in the schema are used in art and architecture. In addition to ice-rays, there are examples of the schema in the modernist paintings of Georges Vantongerloo and Fritz Glarner – perpendicular divisions or nearly perpendicular ones – in the Malagueira housing project of Alvaro Siza – perpendicular divisions again – and in Medieval building plans in Venice – neat tripartite divisions with T-cuts or parallel ones. Even Frank Gehry uses the schema in his fancy roofs – so much for originality! But really, all of these examples are distinct. There's no mistaking one for another.

I can always expand a schema by adding copies of it, where $t(x) \rightarrow t(y)$ is a copy of the schema $x \rightarrow y$ under a transformation t. (A schema and its copies are equal, because rules apply under transformations.) This lets me find similar parts of a shape in parallel and change them in similar ways. For example, if I use a reflection, I can define schemas for bilateral symmetry in classical building plans. Of special note, Palladian villa plans start off with rules to proportion grid cells – they're in this schema

$$x \rightarrow x + x'$$
$$+ t(x) \rightarrow t(x + x')$$
$$\overline{x + t(x) \rightarrow (x + x') + t(x + x')}$$

and in comparable ones. Then rules in the inverses of schemas like this

$$x \rightarrow x' + x''$$
$$+ t(x) \rightarrow t(x' + x'')$$
$$\overline{x + t(x) \rightarrow (x' + x'') + t(x' + x'')}$$

or alternatively,

$$x + t(x) \rightarrow div(x) + t(div(x))$$

concatenate grid cells in room layouts. Once again, there are shared means and distinct ends – the schema $x \rightarrow x + x'$ is in Aalto, Scharoun, Klee, and Duchamp, and the schemas $x \rightarrow x' + x''$ and $x \rightarrow div(x)$ are good for ice-ray lattices, etc. Is there no end of these surprising correspondences? It's neat to see just how far schemas go. And this exercise isn't gratuitous – it builds a repertoire of schemas with workable outcomes that inform seeing and making in education and practice. In fact, this may be an answer to the elusive goal of effective studio teaching in art and design, and a useful basis for open-ended, creative practice, too.

My ability to define new schemas implies new lattices that tie things together in memorable ways. Using some of what I've done up to now with addition gives me the following lattice for transformation rules:

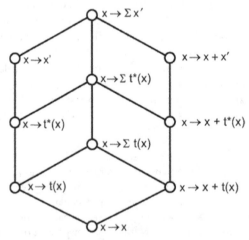

The only new wrinkle is summation. Rules in each of the three summation schemas are good for fractals, fractal-like patterns, and much more – rules for ice-rays are in $x \rightarrow \Sigma x'$. The location of the summation schemas in the lattice shows how rules for fractals include rules for symmetry patterns. (It's fun to notice, too, that any rule $x \rightarrow y$, with shapes x and y in a given algebra, has the form x → Σt(prt(x)), or its inverse. There are two cases: prt(x) is the empty shape or a basic element. This gives details that the universal schema x → prt⁻¹(prt(x)) doesn't – at least it involves the first table at the beginning of this essay. How would you use the boundary operator b to define rules with shapes x and y in different algebras to involve the second table?)

A question I'm asked almost as frequently as

Q1 What rule(s) should I use?

is this one:

Q4 What should I do in the studio?

There are many curricular options in my schemas, and in the way they're defined. Most of these are untried and so, full of undimmed promise. This is what I'd do first, although other scenarios are equally plausible, including ones that simply augment the project-oriented studios that are the norm today. To start, I'd teach about embedding and rules, with no math or fancy technology. Drawing with tracing paper is all it takes – trace this and redraw the rest. Repeat this over and over again until you're satisfied, noting each time what you see, A, and what you add, B. That's the rule you've used: A → B. (The part A includes what's left out, but A may be more if it overlaps B.) Don't be afraid to change your mind freely about what you see. Pretend you're looking for the first time – you've found something new – and trace and redraw what you like, without thinking too much about what this should be. Then I'd go through examples of various schemas – how they're defined and how they classify rules, including the special ones you've tried. Schemas are good mnemonics. They're easy to remember because they lack the spatial detail in individual rules or sequences of rules. Schemas can be given in plain words and recalled by name. They let you talk about what you see and do in your own terms, and how this might go on. Schemas can be changed and combined, so that you can cook up the ones you need as you need them. And you're free to try the rules they define on whatever you see. Schemas define rules for different purposes at different scales, with novel results. The division schema x → div(x) is good for vernacular designs in the crafts and for art and architecture, and surely, for city design, as well. (Savannah, Georgia repeats a motif in Chinese window grilles.) Art and design are observational – what you find here may work there when you notice that things look the same or change them to match. The reason for schemas isn't to decide or limit what you can do but to define rules for seeing, so that you're free to go on as you please. Whether I tell you or not, the goal is to learn to calculate with shapes and rules – to use your eyes to see and do more. (A similar task is to get computers to define and use schemas and rules with embedding – anything less is visually incomplete. This may herald full-fledged computer support for art and design.)

Of course, there's also the desire – perhaps it's an obligation – to teach students to be creative. I'm asked about creativity whenever I talk about schemas and studios. This gives me the chance to test the answers I've heard. None of them really works, except one that enjoys broad yet mute approval. It seems that everybody knows how to get something new, but is afraid to let on. In a word, the secret formula is this:

copy

This has to be wrong. At MIT, entering graduate students are given a manual warning them about copying and its perils, albeit this may spoil their chances for success in science, engineering, management, and architecture. (Only the humanities may be immune. Scholars aren't scribes, but how about writers and poets: "poets steal … and good poets make it into something better, or at least something different.") Science depends on repeatable experiments and results. And it's hard trying to imagine any kind of design without some kind of copying – "reverse engineering" and "best practice" are exactly that, best practice. Noticing someone else's rules and checking them out is flagrant copying, and it's learning to use schemas, as well – maybe the very schemas creative artists and designers already share. Is this kind of learning bad? In fact, using a rule A → B to calculate with shapes is copying: trace everything except A and copy B. No copying means no calculating – so much for MIT! The loss is pretty small, given what's at stake. We're admonished again and again not to copy in school and in professional life – it's cheating. That's the final word. But what are you doing if you copy

cheating

letter for letter to see how terrible this is – and h happens to look like r? (The typographical rule

$$_,h \rightarrow {}_,r$$

is in the schema x → prt(x) in my original lattice – or in a more expansive schema, say, x → prt(x)', to include typefaces that are embellished more than Helvetica – and it's far from cheating when I try the rule. John Dee may have enjoyed this. Rules in x → prt(x) neatly parse his hieroglyphic monad with its multivalent significance – h's are charged with meaning, too.) Petty scruples block the way. Don't be squeamish – copying and plagiarism are OK in art and design. T. S. Eliot (above) knows poets "borrow" from poets, and that the risk is worthwhile when "the good poet welds his theft into a whole of feeling which is unique, utterly different from that from which it was torn." Duchamp recommends copying in *The Blind Man*. R. Mutt's *Fountain* is "plagiarism, a plain piece of plumbing." His theft is trivial and "unique, utterly different" –

> Whether Mr. Mutt with his own hands made the fountain or not has no importance. He CHOSE it. He took an ordinary article of life, placed it so that its useful significance disappeared under the new title and point of view – created a new thought for that object.

The idea is to let embedding work for you – to see things in new ways, and to have the means to do something about this without having to invent them, as well. That's what schemas and rules are meant for – to reconfigure what you see as you go on – and trying them is creative. In logic, linguistics, and computer science, creativity is just recursion and identity. (Chomsky is famous for applying this equivalence to language.) This is a pretty good definition – in fact, the best I've found. It's probably a little technical, but at least I know what it means – it's not framed in the usual way in elusive terms that beg for more explanation. And it's easy to locate the source of originality and novelty. But I also think this definition is hopelessly incomplete, at least visually. Without embedding, it doesn't go very far. New things get more and more complex, adding the same old building blocks (units) to the same old building blocks, etc. It's counting higher and higher, with no chance to see what's going on and how this alters in surprising ways. Creativity takes recursion and embedding. Embedding lets me copy without copying by rote – there are no building blocks – in a kind of visual improvisation that's full of surprises. Anything can change anytime. This helps to explain why art is art. And it may explain where art comes from, too. Origins are apt to be confused and obscure, yet embedding and shapes cut through the mist with a blunt irony. The origin of art is *clear* only because seeing is *vague* and *ambiguous*. There's art anytime I observe the miscellaneous things around me and change them to see as I please. For example, if figures, formulas, schemas, and rules happen to look like section headings – and why not – then

What next?

My answer to this question is pretty much the same as my answer to Q1 – anything you want. I've already talked a lot about schemas and rules, and there's a lot more to say about them – about how they're used and what they do when I calculate with shapes. In every direction I look, there are new and useful things to explore. Nonetheless, the question has an established provenance that decides the way to go. This locus adds more about art and design, and how they fit together with calculating.

The question is Alberti's motto – QVID TUM? Lionel March shows that its seven letters have a coded numerical meaning – the entire expression yields the perfect number 28, that's the sum of its proper divisors. Alberti put art and mathematics together in a single stroke nearly six centuries ago. And shape grammars try this today in another way, when they meld recursion and embedding to calculate by seeing. Alberti does a lot with painting, sculpture, and architecture, and their mathematics. There's recursion – in *On the Art of Building in Ten Books* in which cities, buildings, and rooms are compared in a neat passage on compartition – and symmetry, proportion, and perspective. All of this is the stuff of shape grammars. Recursion animates rules. And rules in schemas, for example, $x \rightarrow x + t(x)$ and its several variants in my lattice of schemas for transformation rules, handle symmetry, proportion, and perspective. This may be noteworthy – it's all very easy – but it's not too surprising. Ho hum, you can calculate in shape grammars with some familiar mathematics. That's exactly what shape grammars are for. But then Alberti anticipates the twin tables at the beginning of this essay for basic elements in shapes, and their dimensions and boundaries, and this coincidence may tie to shape grammars in a deeper and more exciting way.

Alberti starts *Elements of Painting* with basic elements – points, lines, planes, and solids, or bodies, as he calls them – according to Euclid's *Elements* (cf. my first table):

1. The point is said to be that which cannot be divided at all into any parts.

2. A line is said to be [like] a point stretched out in length. Thus the longitude [length] of a line can be divided, [but] not its latitude [width].

3. A surface is said to be produced almost as though extending the line width-wise, and thus its length [and width] can be divided, [but] not its depth.

4. A body is said to be whatever can be divided by length, by width, and also by depth.

Alberti then goes on to add an original description of basic elements and their boundaries – the similarity to shapes is uncanny (cf. my second table):

1. We call a body that which is covered by surfaces, on which our vision comes to rest.

2. We call surface that extreme skin of the body, which is encircled by its edge.

3. We call edge the whole circuit, almost [like] borders, where the suface ends, which place I call *discrimen*, a term taken from the Latins.

4. A *discrimen* proper is the line drawn from the forehead that divides the hair so that some falls to one side of it and some to the other. Thus using this simile, in our case *discrimen* will be that length in the middle of two surfaces that divides one from the other, and this length terminates in two points.

5. I call point that extremity where several lengths and *discrimens* share a common termination.

And Alberti may know that basic elements of dimension greater than zero can be divided anywhere and that they don't exceed finite bounds. Aristotle frames the same idea for continuous magnitudes:

A continuous magnitude can be divided beyond any given smallness, but it cannot be increased above any given greatness.

All of this taken together seems enticingly close to embedding for basic elements, and shapes and their boundaries. Does Alberti, adding to Euclid and Aristotle, have seeing in mind? I'd like to think so – Alberti is pretty good company – but it's a real stretch. There's nothing to suggest calculating with shapes, or that parts of shapes can be cut out and traced in alternative ways. The text I've quoted is more of a pro forma preface to measurement and the mechanics of drawing than anything else. Alberti's art and mathematics seem mostly to delineate geometry and transformations. The Boolean half of shape grammars is incomplete, its outline visible merely here and there. It depends on embedding in my algebras of shapes, and begs for a full explanation and examples of how it works. But I'm prone to wishful thinking – I'm used to seeing whatever I want to and have no qualms about it. Maybe I should try and look again – maybe in a new place. It's easy to be surprised.

The parts of Alberti that impress me most are more directly about art and design, and may appear at first to have little or no mathematical content at all. Alberti's bold account of art and its origin is especially intriguing. It makes embedding unavoidable – it's pure shape grammars with schemas and rules. Alberti sets things up in the beginning few lines of *De Statua*, sometime around the middle third of the fifteenth century, when he considers how we learn "to create images and likenesses." It's in this marvelous passage:

I believe that the arts of those who attempt to create images and likenesses from bodies produced by Nature, originated in the following way. They probably occasionally observed in a tree-trunk or clod of earth and other similar inanimate objects certain outlines in which, with slight alterations, something very similar to the real faces of Nature was represented. They began, therefore, by diligently observing and studying such things, to try to see whether they could not add, take away or otherwise supply whatever seemed lacking to effect and complete the true likeness. So by correcting and refining the lines and surfaces as the particular object required, they achieved their intention and at the same time experience pleasure in doing so. Not surprisingly man's studies in creating likenesses eventually arrived at the stage where, even when they found no assistance of half-formed images in the material to hand, they were still able to make the likeness they wished.

Is this something to take at face value? To be sure, it's easy to find the faces of superstars and deities in strange places – there are many curiosities for sale on eBay. Nonetheless, this is scarcely a start. Alice's Cheshire cat smiles at her from its perch in a tree whenever she blinks. Big dogs and wild things are everywhere you look. And the sky is full of novel creatures on land and sea, day and night. But can I do more with these fanciful images? How are they resolved? How do they change? Are they useful in art and design? Maybe I can do what Alberti suggests with my schemas for rules. It's no big deal to trace a contour or fill in an area "in a tree-trunk or clod of earth" with rules in the inverse of the schema for parts

$$x \rightarrow prt^{-1}(x)$$

Or I can fill in a contour whether it's complete or not using rules in the schema

$$x \rightarrow x + b^{-1}(prt^{-1}(x))$$

This is easy to get adding the identity $x \rightarrow x$ to the schema $x \rightarrow b^{-1}(prt^{-1}(x))$ that relates point and line to plane, etc. And if the contour merely evokes a fragment of what I'd like to see, then I can try something more ambitious to supply whatever seems lacking to achieve a true likeness

$$x \rightarrow x + prt^{-1}(b^{-1}(prt^{-1}(x)))$$

But this schema may be unnecessarily abstruse with too many inverses and nested parentheses. I guess a little calculating is called for, with distinct rules and schemas – then again, my schema equals this one

$$prt(b(x)) \rightarrow prt(b(x)) + prt^{-1}(x)$$

that may be easier to grasp. In the reverse way, I can outline areas by adding the identity to my schema for random patches in clouds

$$x \rightarrow x + b(t(prt^{-1}(x)))$$

and letting t be the identity transformation. And if bits of what I see are wrong, then I can take them away and mark the part that seems right to me now

$$x \rightarrow prt(x) + b(prt(x))$$

Meaning is mine to make, as I go on in this open-ended process. Whatever rule I try in any schema lets me see more in a new way. And I'm free to look again with the same rule or another one.

(This proves copying is creating. That copying is cheating is an axiom at MIT. But cheating contains creating because h contains r. So copying is creating, too. Alberti would likely agree: "slight alterations" go two ways

cheating $\Leftrightarrow$ creating

I can "add [and] take away ... to effect and complete the true [sic] likeness." Simply try the rule

$$_+ h \rightarrow {}_+ r$$

in the schema x → prt(x), or the inverse of the rule. An eight-letter word is in an eight-letter word, letter for letter. Odd – but not difficult to check by means of embedding, if I type both words with equal spacing and place one over the other, so that cheating conceals creating. How will I ever decide which word to read – the one on top? I guess there really are no non sequiturs.)

Alberti finds various things in tree-trunks and clods of earth, and his extravagant way of seeing (embedding, tracing, copying) extends easily to drawings, paintings, photographs, sculpture, buildings, "ordinary articles of life," etc. They're sources for original art and design, too. And equally, I can go on in myriad ways with the rules in my schemas. I can see as I choose – no-holds-barred. There's embedding everywhere I turn, every time I try a rule. Calculating by seeing makes a real difference. First, embedding gives me a way to go on "by diligently observing and studying," not knowing in advance what I might find. I'm free to change my mind as I calculate with shapes. I'm the one to decide how I see – my experience is mine alone. Then, ready access to schemas and rules lets me interact with shapes continuously with no perceptible break or pause – it's being able to use schemas and rules when I choose, to see what I want and to do what I want, with novel results. The schemas I've got are amiss if I can't use the rules they define as freely as I like. Embedding ensures that my effort isn't wasted. Rules work in just the right way – I only have to look!

I'm endlessly amazed at how straightforward it is to find overt examples of embedding in art and design, and at how often embedding is rediscovered and described anew – even if this is as far as it goes. The various lists I make seem comically haphazard because of everything they miss. I always start a new list with rules – embedding is used whenever a rule is tried. Then several neat examples come reliably into view. Art historians must know that Ernst Gombrich and Michael Baxandall depend on embedding to ground their work, as "projection" – perception à la Rorschach – and "cognitive style." In fact, embedding appears to be intentional across-the-board in art – from Alexander Cozens's "blotting" that extends Leonardo da Vinci's use of chance images in design, to Picasso's *Bull's Head* (bicycle saddle and handlebars) and *Baboon and Young Vallauris* (toy cars), to Salvador Dali's grandiose "paranoiac-critical method" (the tendency to see things in diverse ways makes paranoia grow – a shape grammar is every paranoid's dream). Donald Schon finds "reframing" and "displacement" in design and like professions – management, planning, psychotherapy, etc. – Ellen Langer roots seeing and other creative experience in "sideways learning" and "mindfulness," and David Perkins implicates "noticing" and "contrary recognition." (Contrary recognition implies that seeing is merely routine – how else can seeing be wrong? Remember what you're trained to see, and mistrust your eyes when you look again. Embedding is too lax – identity is enough – and being creative seems hard. Seeing as if for the first time takes a kind of "negative capability." Langer shows this to be vital in learning.) Aptly, William James describes reasoning as "learning and sagacity" – the latter is explicitly embedding – and relates both to creativity. Creativity is what teaching should foster and what the studio is for – learning schemas and using the different rules they define with sagacity. Like themes run through art education. In a welcome retreat from "rationality," Elliot Eisner stresses "flexible purposing" (he thanks John Dewey for the term) and visual improvisation – redirecting your work and redefining your goals, as options emerge in a stream of surprises. (Lois Hetland and others at Harvard's Project Zero find this in high school art studios. Herbert Simon describes painting and social planning in the same way.) Do surprises comprise a kind of creative consciousness? The list goes on. Ambiguous figures turn up in Neolithic times. The classical Greeks reject checkerboard patterns in weaving because they're visually unstable and so, "barbaric" – that's it for perceptual rivalry. Hieronymus Bosch paints topsy-turvy pictures in which oarsmen in the sky steer boats to clouds at sea. Anatomical drawings – intentional or not – appear in Michelangelo's frescoes in the Sistine Chapel. (Would Alberti or Leonardo or Dali object to this? Art historians today may not be as generous.) Magic viewing stones abound – Chinese scholar's rocks, Japanese *suiseki*, and Korean *suseok*. And in *Mother India*, M. F. Husain traces the contours of a naked image in a map of India. (Images and maps may be categorically distinct. This adds a logical twist to the political uproar over *Mother India* – an image that can change is part of a map that can't. Slick trick! I wonder if the logic police will notice any inconsistency. But this hardly matters: Husain fled Mumbai because of creative excess and a lapse from grace.) These are a few easy examples of seeing – opening your eyes to reconfigure things freely in new ways, and giving up rote norms and prior expectations (aka standards) to make use of what you find. Creative experience isn't routine – it exceeds what memory holds and training requires. Schemas and rules exploit this with embedding in art and design.

Alberti linked mathematics, and art and design – using the one in the other two. The *Nexus Network Journal* website describes this relationship for architecture:

> There are many connections between architecture and mathematics: mathematic principles may be used as a basis for an architectural design, or as a tool for analyzing an existing monument; architecture may be a concrete expression of mathematical ideas, becoming, in a sense, "visual mathematics."

Art and mathematics is putting mathematical results in the hands of artists or designers to make something that's useful or worth looking at, or to describe finished work. Art and design aren't mathematics – they're fine without it, but they use it when it helps to go on, in the same way they use everything else to add something extra. Ideally, art and mathematics enjoy the same kind of relationship that holds for physics and mathematics, although physics may have a stronger and more intimate tie. It's hard to imagine physics without math, and there's a kind of reciprocity – they enrich each other. Nonetheless, mathematics works in art and design. Schemas and rules in shape grammars are useful. Shape grammars do Alberti's mathematics – and then they do more. There's a marvelous twist that makes them inevitable.

Shape grammars turn art and design into mathematics – well, at least calculating – by showing that rules in schemas match what's tried in practice. In fact, you can't avoid using schemas and rules. There's calculating anytime and anywhere you look – tacit and not. That's the gist of my formula

design = calculating

The trick is to be aware of the equivalence, so that you can use it effectively in teaching and practice. The equivalence goes two ways – there are big gains on both sides of the equal sign. Art and design enlarge calculating with embedding, while calculating with schemas and rules explains how seeing and doing work in art and design. Embedding is the key. And ironically, Alberti grasped this, too, finding parts of images and likenesses "in a tree-trunk or clod of earth," and then "correcting and refining the lines and surfaces" to get what he wanted. This is seeing and doing with shapes, and it returns full circle to my original question:

Q1 What rule(s) should I use?

With schemas and embedding, my answer is never in doubt:

A1 Use any rule(s) you want, whenever you want to.

Schemas ensure that I always have a repertoire of meaningful rules to try, and embedding ensures that whatever rule I use works to surprise me, in an open-ended process. I can always go on to something new. This is calculating by seeing with recursion and embedding. Yes – it's art and design.

Background

My shape grammars extend Turing machines – shapes, embedding, and continuous transformations allow for more than symbols, identity, and discrete translations. What I see is too fickle to pin down in letters and words. There's no deciding what parts shapes have before rules are tried. Basic elements are continuous – lines, planes, and solids divide freely, so that units fail in art and design. Shape grammars work outside of Turing's logic and combinatorics with their all-or-none bits (0's and 1's). Others also find this way of calculating too limited. John von Neumann would like to calculate with real and complex numbers. Lenore Blum and her collaborators pull this off in numerical analysis and scientific computing. Then Newton's method to solve polynomial equations shows how to calculate, not Turing machines. Turing's bits and codes won't do in processes of this kind. Bits and codes aren't real and complex numbers, the same as symbols aren't shapes. Continuous mathematics (analysis) – like seeing – needs a new way to calculate that isn't combinatory. Science, and art and design are largely two cultures. But when it comes to calculating, there's an amazing convergence. Blum's way of calculating and shape grammars coincide – Turing machines are a special case in both. And there may be more to compare – perhaps calculating over rings, even if this seems to be better suited for numbers than shapes that are more intuitively described in terms of tracing paper, and drawing as a trained draftsman (embedding and maximal elements). Whether any of this is of use technically doesn't matter. It strikes an uncharted arc from science to art and design that proves they're equal, as it bolsters their separate ways of calculating. But shape grammars aren't simply math and rules to calculate. That's far too one-sided and abstract. Shapes are concrete, and rules are inherently phenomenal when they're tried. Shape grammars deal with the vagaries of visual experience in the real world.

H. W. Janson deftly traces chance images (embedding) in Alberti, Leonardo da Vinci, Cozens, and others in "The 'Image Made by Chance' in Renaissance Thought." Kim Williams, Lionel March, and Stephen Wassell provide the English for Alberti's *Elementi di Pittura*. I read Cecil Grayson's translation of Alberti's *De Statua* while I was learning about checkerboards and why the classical Greeks shunned them, in Wolfgang Metzger's Gestalt tract, *Laws of Seeing*. (I like to read unrelated things side by side at the same time, to see if they overlap in any other ways. It's uncanny how seemingly distant viewpoints augment one another.) *Mother India* highlights the contrast between images and maps in Doxiadis and Papadimitriou's *Logicomix* – the exact position is on page 211 with grid coordinates A3. A speech bubble points to an image of Christos wandering around, hopelessly lost in Athens:

YOU SEE,
IMAGES CHANGE,
BUT **MAPS** STAY
THE SAME!

The classical Greeks would have liked Christos's maps – in philosophy (Plato, *The Republic*, Book VI) and in weaving. But what are maps? How do you get one? What do they cost? Does a map show where to go? Philosophers may know the way – they aren't "blind" like the rest of us, not knowing ahead of time what to see, but have "perfect vision" of the abstract world that's an eternal guide. Maps are "clear patterns" (structures) in the mind's eye or computer memory that can't be seen beside tree-trunks and clods of earth. Otherwise, maps are simply checkerboard patterns (shapes) – grids with precincts in red and black, or any contrasting colors (are these bits?) – that can alter erratically. It's farewell to weaving when nothing stays the same. Are philosophers afraid of this instability? Does it scare artists and designers in the same way? M. F. Husain opts to see rather than settle on what's fixed. Seeing isn't routine when you use your eyes in the concrete world, and aren't obliged to recite by rote the "eternal and unchangeable" things that reason sanctions in philosophy and weaving. Art and design just go on. They're free to "wander in the region of the many and variable" – no end in sight.

A comprehensive discussion of shapes, embedding, and rules and schemas is in my book *Shape: Talking about Seeing and Doing*. It begins with Plato, and wanders around to calculate in algebras of various kinds. For rules and schemas, the key idea is this: the descriptions given to define a rule in a schema needn't apply to any shape produced from the rule. Descriptions are useful but aren't fixed permanently. What I say I see can change freely as I calculate. For example, the schema

$$x \rightarrow x + t(x)$$

defines the rule

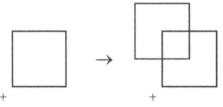

when its right side is two squares of the same size – x in the left side of the rule, and x plus t(x) in the right side. The rule applies to the square

to produce the shape

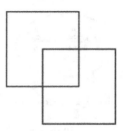

with many surprises. The shape needn't be two squares – maybe it includes a pair of L-shaped, concave hexagons that share two vertices (schemas with nonempty left sides x define rules for these hexagons and any other parts), or maybe the rule is used again to add a square to the small, central one

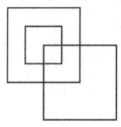

Then the two squares in the shape

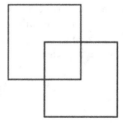

are a square and a concave octagon. There's no reason to keep track of the figures that make up a shape. They disappear without a trace when they combine. Parts alter erratically – I can try any rule I want anywhere I want. And there's more to see adding various transformation rules in the schema

$$x \rightarrow t(x)$$

Just the rotation rule

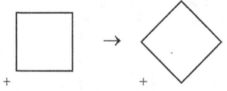

livens things up – there's even some new mathematics with palindromes, when I try (and fail soon enough because there's too much going on all at once) to parse the sides of squares into meaningful pieces of various lengths, that are related in a definite sequence. And translation rules like this one

+   +

that move squares wherever you want in Cartesian fashion – left or right and then up or down – add to a wonderful variety of nifty changes that are sure to exceed expectations for squares

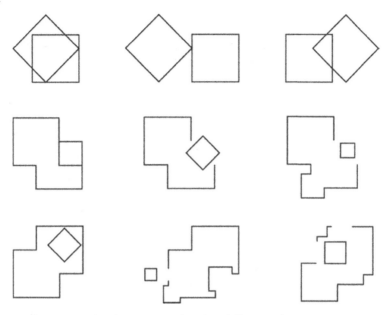

It doesn't take very much rule-wise to make a big difference shape-wise. The leverage can be remarkable. Identities in the schema

$$x \rightarrow x$$

are enough to show what I mean. But all of this is already in my lattice of schemas for transformation rules. I can follow it through going back and forth between top and bottom, taking various paths. This isn't traversing a classical building from place to place in a rhetorical manner. Nonetheless, my lattice is a pretty good mnemonic that leads to something new. Whoever said art and design aren't calculating? It's no use trying to keep them apart. Calculating goes on and on, doing exactly what I see.

The schemas I've described in this essay are generic. This makes them memorable, and easy to try in novel ways. Schemas are for teaching – there's learning and sagacity. But there are individual stylistic distinctions and other fine points that need to be explained in practice. For example, when individual artists and designers use the division schema $x \rightarrow div(x)$, there are distinct results. These differences depend on the assignments given to apply schemas to shapes to calculate. Assignments satisfy predicates when they associate shapes with the variables in schemas. And predicates may restrict how operators are used. The operator $prt(x)$ and its inverse are pretty general and may be productively reined in with predicates – perhaps $prt(x)$ picks out polygons or parts that are maximally connected. Assignments and predicates involve an added layer of technical description in words and formulas, that's in my book, too. And notice how schemas and assignments recall the ancient distinction for unity and variety – there are many versions, including Alberti's. In my terms, there's unity in a shared repertoire of schemas, and variety in the assignments to try them. Relationships for schemas and assignments, like unity and variety, have yet to be examined in detail, but they're ripe for empirical study and can be tested by calculating. For example, what angles work in ice-rays to limit how divisions are made? Are other angles worth a try? Is there more to ice-rays? This adds a new technique to reinforce known classifications in art and design, and to suggest fresh alternatives.

In one way, schemas and rules go beyond what Alberti recommends. Baxandall notes that Alberti was the first to consider pictures in terms of *composizione* – a hierarchy of independent parts that's the same for artist and viewer alike, allowing for a kind of communication between them. Alberti took the idea from classical literary criticism where *compositio* describes the structure of sentences in four tiers. A picture is divided into bodies that are divided into members that are made up of plane surfaces. There's making and seeing going up and down the hierarchy to meet narrative goals. Alberti's use of hierarchy in pictorial composition has been adapted in many ways in art and design – J.-N.-L. Durand's system of architectural composition and Alexander's architectural "pattern language" are two instances. But today, the "principle of compositionality" is Frege's in logic – what an expression means depends on the meaning of its parts and how they go together. This informs linguistics and computer science, too, and it's evident in engineering and biology, especially in design education and practice (for example, go to http://biobricks.org for the use of "*BioBrick* standard DNA parts that encode basic biological functions"). The key idea seems to be inevitable: unit parts with predefined functions combine in compounds with derived functions, and this repeats in a hierarchical ascent, combining compounds and functions to complete a design. Are schemas and rules compositional? The use of operators and inverses, composition, and addition to define schemas has a full-fledged compositional flavor. This may explain some of its allure – it's really neat the way everything fits together in useful schemas. And the rules in schemas are compositional if you choose to use them in this way to calculate with shapes. Even so, rules do far more that's not compositional. Embedding is taking a chance on what to see. There's a new way to go on every time you open your eyes and look again. Parts alter as rules are tried – parts aren't permanent and independent when they combine, but evanescent, fusing and dividing erratically. Putting things together, so that they're fixed in known relationships, may not explain what's there, or even begin to. (This goes for perception and physics alike – the easy congruence between Robert Laughlin's "emergence" and my embedding proves in another way that science, and art and design match. But the equality fails for "reductionism." This provides a compositional alternative to emergence with units and fundamental laws: then embedding is identity à la Turing. Next to reductive science, art and design hold sway. This is good news for art education. Schools can teach art without any qualms – and not because art is instrumental for basic subjects or something to appreciate. The lessons of embedding/emergence unfold when art is art. There are untold questions to ask and answer. For starters, are unit parts emergent or not? Are they fixed once and for all? But physicists aren't worried about art education when there's more science to do. Laughlin is keen to dismiss reports that science is dead with new observations and laws. A reductive theory of everything is a tempting illusion that's hard to deny. Nonetheless, it's folly – discovery is open-ended. Embedding/emergence is why in science, and in art and design.

Laughlin highlights this parallel. Maybe art is physics, and vice versa – they have much in common. "How very interesting.")

Hierarchies feel safe, but they're not – they collapse suddenly as parts change, in one compositional catastrophe after another. It's like having the rug pulled out from under you when you least expect it. Everything is topsy-turvy. It's easy to be afraid or angry – you worked hard to figure out what was going on, and to make it all fit just right. Still, the surprise of a fresh start can be exhilarating – you've created (found) something new. The rule

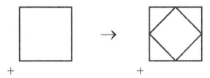

in the schema

$$x \rightarrow x + t(x)$$

combines a pair of squares, that the erasing rule

in the schema

$$x \rightarrow$$

divides into a quartet of triangles. But how does this work compositionally, if I start with squares? Aren't they fixed once and for all? Where do the triangles come from? And where do the squares go? Is there anything in between squares and triangles that's something to see? Parts don't appear and disappear erratically, without rhyme or reason. What happens to meaning then? It seems simple enough to use one rule or the other in the obvious way to define separate hierarchies for the shape

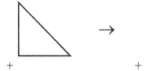

But the hierarchies are evidently incompatible – categorically and numerically distinct – because my rules resolve unit parts. What's going on? Do different hierarchies mean different shapes? Let's count their parts by kind to find out. It's perfectly clear: no one believes that two independent squares

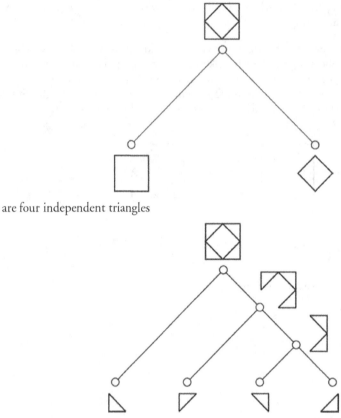

are four independent triangles

But this is merely counting. I see four triangles when two squares fuse and divide. Can I show how this goes? Perhaps in retrospect – guessing what I'll see before I look won't work: triangles may be pentagons or various other shapes. But once I've seen squares and triangles, I know what to do. First I need to match these figures, in the way they're combined, to define the pieces they have in common – I guess units aren't units after all:

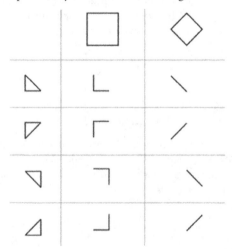

Then I can add these parts of squares and triangles to my original hierarchies to divide shapes in finer detail. The results join across a frontier of common parts in a non-hierarchical network with confusing crossovers:

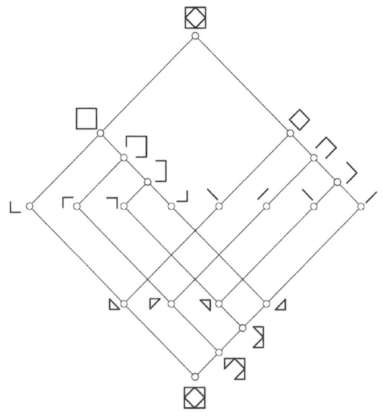

Everything is something else. Going from top to bottom in the network, the four corners of the outside square switch to four separate triangles, while the four sides of the inside square switch in this way, too. In the reverse direction, the right-angled corners of the four triangles make the large square, and the sides of the triangles combine in the small square. Triangles are all alike, with a right angle and a hypotenuse, but neither three sides, as I learned in school, nor three angles, as the word implies – so much for stock definitions. Yet squares divide in this way – they come in two kinds, with four sides or four right-angled corners. Aren't squares all the same – maybe not? How many kinds of squares are there? (Everything is perfectly clear – I embed distinct figures, possibly with alternative structures, or do these figures and structures emerge in a fit of self-organization?)

This isn't compositional: then triangles aren't mixed up in squares. But this is how it goes in art and design, not just once or twice with squares and triangles, but again and again in an open-ended process with whatever it is I happen to see – technically, every time I try a rule. It's worth repeating: parts can change freely anytime with no way of knowing in advance what these changes might be. I have to wait and see. I may divide the shape into three kinds of pentagons or congruent hexagons, or some number of big K's and little k's. (Mathematics is never far away. Big K's and little k's fall into a dense series κ of dense series, each with a leading K and successive k's with growing stems. Every series in κ is a proper initial segment of every series that follows it, under a similarity transformation that fixes the length of the segment and where little k's are alike.) Where does all of this come from? Alberti's *composizione* and Frege's principle of compositionality are simply too narrow and inflexible to account for everything there is to see. They play it safe, and miss what's salient next. (I sometimes think that museums also err in exactly the same way, with their "self-guided [sic] audio tours." Words may keep you from using your eyes when an "expert" reminds you of what to see. This fails, especially in education, because embedding is involved.) But composition is too elegant not to give it a serious try – at least it shows you understand what you're doing. Hierarchies are as cognitively seductive as anything I know – it feels good when things fit and mean what you want them to – and hierarchies help to communicate with others. Nonetheless, how do you define a hierarchy before meaningful parts are resolved and fall into place? How many times do you change your mind about what you see, as you wait for something to gel? What makes this possible? Are there reasons to look again? Then, QVID TUM? That's why there are schemas and rules – they're not for composition but to look again. There's nothing to put together, only new things to see. Seeing (perception) precedes understanding and communication. You can't get around it – schemas and rules are needed to define hierarchies. They do the trick and a lot more with embedding. There's an alternative principle in this for art and design that's pre-compositional – the parts you see and use can vary freely and needn't include any of the parts you've combined. There's plenty to do before meaning sticks, and this always starts over when you go on. This is negative capability, indeed.

Composizione has an unruly twin – *varietà*. In Baxandall's formula, "*Composizione* disciplines *varietà* : *varietà* nourishes *composizione*." James Gips and I describe this kind of reciprocal relationship for unity (hierarchy) and variety in terms of George Birkhoff's aesthetic measure $M = O/C$ and our own information-theoretic measure E_Z, in our book *Algorithmic Aesthetics*. My current thoughts on such measures can wait for another time. At the very least, something more dynamic may be necessary. Both Birkhoff's aesthetic measure and E_Z fix aesthetic values once and for all, but this needn't be the case for E_Z. The relationship between schemas and assignments as unity and variety raises many fascinating questions, too. A given repertoire of schemas and various assignments may explain surprising similarities among diverse artists and designers. Baxandall lauds Landino's use of *compositione et varietà* to compare Filippo and Donatello in precisely this way. This is "one of the prime bearings of Quattrocento cognition." But for schemas and assignments, there's a crucial shift from cognition to perception, and escape from the stringencies of hierarchy to the freedom of embedding. Every time I decide to give up on an idea – this time, unity and variety – it strikes me again in a new way. Maybe ideas are more like shapes than I thought. It's good to keep them around – you never know what you might see next.

My schemas define rules that produce shapes by calculating. This departs somewhat from the way schemas are ordinarily described and used in art and design, but it may include such schemas, as well. Goethe's *Urpflanze* is the acknowledged paradigm for schemas and how they're used: it provides a comprehensive survey of plants that are elaborated individually when rules are applied to permute and transform organs of specific kinds. Then Gottfried Semper's *Urhutte* is an archetype for buildings that's framed in the same way. And perhaps the shape

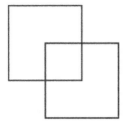

is another example that's easy to vary with transformation rules in the schema x → t(x). A few of these variations are shown above. But this may be too general, because the parts of the shape aren't resolved and classified before rules are tried – is it two squares, a square and an octagon, twin hexagons that touch twice, or something else? Labels and like devices are readily available for this purpose whenever they're needed. Shapes and rules may also have the relationship for schema and correction that Gombrich recommends. This is marvelously close to Alberti – schemas are outlines in tree-trunks or clods of earth, and rules are for correcting and refining lines (boundaries) and surfaces (areas) to create various images and likenesses. The relationship between schemas as archetypal shapes, and rules to change them to meet the ongoing demands of experience and circumstance seems to be a very common idea in art and design that works with shapes, and the rules I define in my schemas to calculate.

In architecture, and sometimes in philosophy, a distinction is made between schemas (good) and rules (bad) to contrast a synoptic view of possibilities, and combining building blocks or unit parts in *BioBrick, Lego,* or *Tinkertoy* fashion. Semper's *Urhutte* is an ample summary of buildings, while Durand's system is combinatory with its building "elements" (vocabulary) and rules of composition to put them together (although at times, Durand seems to be a draftsman using embedding). But the distinction may not be that clear. A schema has its combinatory aspects, too – Goethe needs permutations and transformations to vary the organs of his *Urpflanze*. And a combinatory system with its vocabulary and rules provides a synoptic view of possibilities that unfolds in a recursive process. Maybe schemas and rules are parallel ways of doing the same kinds of things, with the emphasis placed either on similarities in origin and classification (schemas) – for example, this might be the case for a theme and its variations – or on differences in process and elaboration (rules) – as in a language of designs. Using schemas or rules seems to be largely a matter of taste, or perhaps it's just a matter of words that are so intertwined that there's no use trying to untangle subtleties in meaning – it's better to feel your way or to go by sight and sound. And how I use schemas to define rules, and rules (inverses, composition, and addition) to define schemas may imply another equivalence with new interconnections. What about my previous example? Is the shape

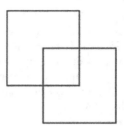

a schema, or is it produced when two squares are combined according to an addition rule in the schema $x \rightarrow x + t(x)$? Whatever differences there may be don't amount to much – changing a schema and combining shapes lead to the same things. In fact, the schema $x \rightarrow t(x)$ is used to define $x \rightarrow x + t(x)$, and is a key part of it. The differences between schemas and rules are trivial, as long as I'm free to describe what I see anyway I want. That's the problem with organs and building elements when they're fixed before rules are tried. And it's the same problem for Goethe and Semper, and Durand – how do their schemas and rules allow for this kind of freedom? Without embedding, neither schemas nor rules let me see as I choose. Limits on what there is to see are set once and for all before I have a chance to look. Seeing is humdrum and routine. This does precious little for architecture, and no more for the rest of art and design. Philosophy may have other aims and goals in mind.

The way I handle schemas in this essay subsumes the schemas in my book, and opens new vistas to explore – especially, in education and practice. Artists and designers are, no doubt, creative. I've been told that this makes them different. (Like views are spread widely. In *The Big Noise*, Oliver Hardy tells Stan Laurel that inventors are "twisted" and uses a manual gesture to show how. There's little doubt when you see Ollie rotate his hand.) But this is no reason to believe that art and design can't be taught – perhaps differently. There are no limits on creativity using rules defined in schemas, when this involves embedding. (At the end of a lecture not too long ago, I was asked why no one is using shape grammars to teach in the studio. This is one of those silly questions that kids ask – "If you're so smart, why aren't you rich?" – that's supposed to make you stop and think. Well, I didn't have to stop and think, I already knew the answer – no one will let me. That's OK. Moses didn't get to the Promised Land. This may have been God's way of thanking him, by sparing him the embarrassment of not knowing what to do once he got there – seeing the present in the way you remember the past isn't the only way to go on to something new – or the opprobrium of running amok trying to get things done too quickly by fiat and not having the patience, after waiting so long, to give others the chance to learn at their own pace. Maybe … but it's still more than a little frustrating to know what rules in schemas can do when there's embedding and not to see rules used in practice. It's high time for someone to let someone – I'm not volunteering – try schemas and rules in the studio, not just in an occasional exercise with an apologetic preamble or in elective studios that aren't taken seriously, but everyday throughout the entire curriculum. Half measures, with the option to abandon schemas and rules when they begin to work or because you're afraid they might, simply won't do the job. It may be daring, but there's little to lose. Results in the studio seem to miss the mark – ask students what they learn, and what they think of this. Why not give shape grammars a serious try?)

Acknowledgments

Thanks to Lionel March for pointing me to classical texts and commentaries, and for taking the time to discuss Alberti, especially the successive sections I quote from *Elements of Painting*. To amuse Lionel, I've given 28 additional references – what next? Catherine Stiny did the speech bubble in the "Background" section with "images" in bold, and "maps" in bold and italic. Maps (symbols) are stressed not once but twice – probably because it's hard to tell them apart from images (shapes). The confusion is ruinous for maps, but merely another way for images to change – hence the double emphasis and warning. It may be too late. The Greeks couldn't save checkerboard patterns in weaving. Maps look the same, and they're equally at risk, even in today's digital world. That's it for maps – shape grammars forsake stability, routine, and rigorous control to meld seeing and calculating.

References

ALBERTI, Leon Battista. 1972. *On Painting and Sculpture. The Latin Texts of* De Pictura *and* De Statua. Cecil Grayson, ed. & trans. London: Phaidon.

———. 1988. *On the Art of Building in Ten Books*. J. Rykwert, N. Leach, R. Tavernor, trans. Cambridge, MA: The MIT Press.

———. 2010. *Elements of Painting*. Pp. 142-152 in *The Mathematical Works of Leon Battista Alberti*. Kim Williams, Lionel March, and Stephen R. Wassell, eds. Basel: Birkhäuser.

ALEXANDER, Christopher. 1979. *The Timeless Way of Building*. New York: Oxford University Press.

BAXANDALL, Michael. 1972. *Painting and Experience in Fifteenth-Century Italy*. Oxford: Oxford University Press.

BIRKHOFF, George D. 1933. *Aesthetic Measure*. Cambridge, MA: Harvard University Press.

BLUM, Lenore, Felipe CUCKER, Michael SHUB and Steve SMALE. 1998. *Complexity and Real Computation*. New York: Springer-Verlag. (Perhaps even mathematicians play with ambiguity – at least in book titles.)

DOXIADIS, Apostolos and Christos H. PAPADIMITRIOU. 2009. *Logicomix: An Epic Search for Truth*. New York: Bloomsbury.

DURAND, Jean-Nicolas-Louis. 2000. *Précis of the Lectures on Architecture*. David Britt, trans. Los Angeles: The Getty Research Institute.

EISNER, Elliot W. 2002. *The Arts and the Creation of Mind*. New Haven CT: Yale University Press.

ELIOT, T. S. 1948. *The Sacred Wood: Essays on Poetry and Criticism*. London: Methuen.

GOMBRICH, E. H. 1960. *Art and Illusion: A Study in the Psychology of Pictorial Representation*. Princeton: Princeton University Press.

HETLAND, Lois, Ellen WINNER, Shirley VEENEMA and Kimberly M. SHERIDAN. 2007. *Studio Thinking: The Real Benefits of Visual Arts Education*. New York: Teachers College Press.

JAMES, William. 1981. *The Principles of Psychology*. Cambridge, MA: Harvard University Press.

JANSON, H. W. 1961. The "Image Made by Chance" in Renaissance Thought. Pp. 254-266 in *Essays in Honor of Erwin Panofsky*. Millard Meiss, ed. New York: New York University Press.

KNIGHT, Terry. Computing with Ambiguity. 2003. *Environment and Planning B: Planning and Design* **30**, 2: 165-180.

LANGER, Ellen J. 1997. *The Power of Mindful Learning*. Reading, MA: Addison-Wesley.

LAUGHLIN, Robert B. 2005. *A Different Universe: Reinventing Physics from the Bottom Down*. New York: Basic Books.

MARCH, Lionel. 1998. *Architectonics of Humanism: Essays on Number in Architecture*. New York: Academy Editions.

METZGER, Wolfgang. 2006. *Laws of Seeing*. Cambridge, MA: The MIT Press.

OPPÉ, A. P. 1954. *Alexander and John Robert Cozens, with a reprint of Alexander Cozens' A New Method of Assisting the Invention in Drawing Original Compositions of Landscape*. Cambridge, MA: Harvard University Press.

PERKINS, David N. 1981. *The Mind's Best Work*. Cambridge, MA: Harvard University Press.

PLATO. 1991. *The Republic: the Complete and Unabridged Jowett Translation*. New York: Vintage Classics.

ROCHE, Henri-Pierre, Beatrice WOOD and Marcel DUCHAMP, eds. 2007. *The Blind Man*. Iowa City: The University of Iowa Libraries. The International DADA Archive. http://sdrc.lib.uiowa.edu/dada/blindman/index.htm.

SCHÖN, Donald A. 1983. *The Reflective Practitioner: How Professionals Think in Action*. New York: Basic Books.

SIMON, Herbert A. 1981. *The Sciences of the Artificial*, 2nd ed. Cambridge, MA: The MIT Press.

STINY, George. 2006. *Shape: Talking about Seeing and Doing*. Cambridge, MA: The MIT Press.

STINY, George and James GIPS. 1978. *Algorithmic Aesthetics: Computer Models for Criticism and Design in the Arts*. Berkeley and Los Angeles: University of California Press. (Don't bother with the library – download a copy of this book for free, at www.algorithmicaesthetics.org)

VON NEUMANN, John. 1951. The General and Logical Theory of Automata. Pp. 1-31 in *Cerebral Mechanisms in Behavior: the Hixon Symposium*. Lloyd A. Jeffress, ed. New York: John Wiley.

About the author

George Stiny is Professor of Design and Computation at the Massachusetts Institute of Technology in Cambridge, Massachusetts. He joined the Department of Architecture in 1996 after sixteen years on the faculty of the University of California, Los Angeles, and currently heads the PhD program in Design and Computation at MIT. Educated at MIT and at UCLA, where he received a PhD in Engineering, Stiny has also taught at the University of Sydney, the Royal College of Art (London), and the Open University. His work on shape and shape grammars is widely known for both its theoretical insights linking seeing and calculating, and its striking applications in design practice, education, and scholarship. Stiny has recently completed a book on design and calculating – *Shape: Talking about Seeing and Doing* (The MIT Press, 2006) and is the author of *Pictorial and Formal Aspects of Shape and Shape Grammars* (Birkhäuser, 1975), and (with James Gips) of *Algorithmic Aesthetics: Computer Models for Criticism and Design in the Arts* (University of California Press, 1978).

Sara Eloy

Instituto Superior Técnico
Universidade Técnica de Lisboa
Av. Rovisco Pais 1,
1049-001 Lisbon PORTUGAL
and
Laboratório Nacional de Engenharia Civil
seloy@civil.ist.utl.pt

José Pinto Duarte*
*Corresponding author

Faculdade de Arquitectura
Universidade Técnica de Lisboa
Rua Sá Nogueira
Pólo Universitário, Alto da Ajuda
1349-055 Lisbon PORTUGAL
jduarte@fa.utl.pt

Keywords: multifamily housing, housing
rehabilitation, shape grammar, design
analysis, algebra, measuring systems,
transformations, space syntax, information
and communication and automation
technologies (ICAT), domotics

Research

A Transformation Grammar for Housing Rehabilitation

Presented at Nexus 2010: Relationships Between
Architecture and Mathematics, Porto, 13-15 June
2010.

Abstract. Existing housing stock must be rehabilitated to meet the new needs of dwellers in the current information society. Consequently, Information and Communications and Automation Technologies (ICAT) must be integrated in living areas. Both shape grammar and space syntax can be used as tools to identify and encode the principles and rules behind the adaptation of existing houses to new requirements. The research proceeds by first identifying the dwellers' demands and determine how the use of technology influences them. The second step is to identify the functional, spatial, and constructive transformations performed manually by human designers in order to infer the corresponding transformation rules and encode them into a grammar. The third step aims to test the grammar in other dwellings that are part of the corpus of the study.

1 Introduction

This paper describes ongoing Ph.D. research that starts from the premise that the future of the real estate market in Portugal will require the rehabilitation of existing residential areas in order to respond to new dwelling requirements, including the incorporation of Information, Communications and Automation Technologies (ICAT).

In an era in which information has a structural role in society, this study proposes to reflect on the transformation of lifestyles that has occurred in recent decades and its impact on the demand for new housing functions and types. The incorporation of new housing functions calls for a new approach to the design of domestic spaces in which conventional spaces need to be complemented with new areas to accommodate activities such as telework and telehealth, in response to the growing demand for access to information and for comfort in homes.

The study focuses on a specific building type, called *rabo-de-bacalhau* (literally, "cod tail") built in Lisbon between 1945 and 1965. This typology was chosen mainly because it is very representative of the period and follows specific generative principles common to all of the buildings within the corpus. Considering that Lisbon has a high percentage of vacant homes and that the existing housing infrastructures are sufficient to respond to the current housing demand in the city, the problem becomes one of how to rehabilitate existing buildings and supply them with features that meet contemporary needs for comfort and access to information, amongst other aspects.

The ongoing research has three objectives:

1. To identify how the use of technology influences lifestyles and creates new dwelling requirements, and how this affects the spatial and functional organization of dwellings. This work complements Pedro's [2000] and Duarte's [2001] frameworks for incorporating new dwelling modes, new domestic groups, and ICAT-related demands. This step has been completed, resulting in the definition of functional programmes suitable for each family profile;

2. To define appropriate ICAT sets to incorporate into the dwelling spaces so as to guarantee environmental sustainability and the social integration of citizens, adapting them to each household according to present and future needs. These ICAT sets are applicable to the individual dwelling as well as the building as a whole, including rehabilitated existing residential stock as well as new buildings. This step has been completed, resulting in the definition of a set of ICAT packages suitable for different family profiles;

3. To define design guidelines and a methodology to support architects involved in the process of adapting existing dwellings and incorporating ICAT technologies, allowing them to balance new dwelling trends with sustainable requirements and economic feasibility.

This paper focuses on the methodology for housing rehabilitation, which is the ultimate goal of the study.

2 A methodology for housing rehabilitation

The ongoing research intends to define a rehabilitation methodology for Lisbon's existing housing stock to enable it to respond to new technology requirements and new lifestyles. The fundamental goal of rehabilitation is to upgrade houses by incorporating and updating ICT and domotics infrastructures, resolving emerging conflicts affecting the use of space prompted by the introduction of new functions associated with such technologies.

To tackle the problem of developing a general methodology for housing rehabilitation we used the *rabo-de-bacalhau* type of building. This allowed us to apply the methodology to actual buildings so that transformation principles could be inferred and then tested. Only with a specific morphology would it be possible to test different hypotheses for functional rehabilitation.

Our work started with the analysis of contemporary demands for dwelling and the development of a knowledge base for the existing ICAT sets for homes, to be taken into account in the application of the rehabilitation methodology. We then proposed a hypothesis for such a methodology, based on the conceptual schema for the design process proposed by Duarte [2001] for the mass customization of housing, following March and Stiny's "Design Machines" [1981]. According to this conceptual schema, the design process consists of two sub-processes: a formulation process that takes user and site data and generates a description of an appropriate house, and a design process that takes such a description and generates a matching solution within a given design language. Accordingly, it was hypothesized that a rehabilitation methodology should encompass four steps, as shown in fig. 1.

The first step consists of gathering the data needed for the rehabilitation process: the household profile and a description of the existing dwelling.

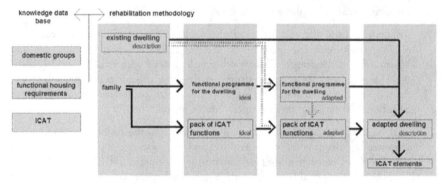

Fig. 1. Basic steps in the planned rehabilitation methodology

In the second step, the household profile is used to determine the ideal functional programme for the dwelling – following Pedro's and Duarte's work on the housing programme mentioned above – as well as the ideal pack of ICAT functions. The functional programme in this case is a description of an ideal housing solution for the family that is not bound by any existing morphological structure or design language.

In the third step, the existing dwelling, the ideal functional programme, and the ideal ICAT pack are used to derive a description of a compromise or adapted solution based on the existing dwelling. Since the solution is influenced by the existing morphological structure, it is necessary to transform the description of the ideal solution obtained in step 2 into the description of an adapted solution.

Finally, the layout of a design solution for the particular family in the particular dwelling is obtained from the description of the adapted dwelling, including the ICAT components needed in the dwelling. In order to incorporate ICAT into the dwellings, two complementary methods were established. Firstly, the introduction of ICAT in dwellings changes some aspects of living (e.g., home cinema, telework, etc.) which were taken into account in defining the functions of contemporary houses. Secondly, the introduction of ICAT in dwellings leads to physical changes in the house due to the need to accommodate cabling infrastructures and terminal elements.

Following the work described above, the methodological hypothesis was to use description grammars, shape grammars, and space syntax as tools for identify and encoding the principles and rules underlying the adaptation of existing houses to meet new requirements. The idea was to use such rules as part of the methodology for the rehabilitation of existing dwellings, as mapped out in fig. 1.

2.1 Shape grammar and space syntax

Shape grammars were invented by Stiny and Gips [1972] more than thirty years ago. They are "algorithmic systems for creating and understanding designs directly through computations with shapes, rather than indirectly through computations with text or symbols" [Knight 2000]. A shape grammar is a set of rules that are applied step-by-step to shapes to generate a language of designs.

Shape grammars are generative because they can be used to synthesize new designs in the language, descriptive because they provide for ways of explaining the formal structure of the designs that are generated, and analytical because they can be used to tell whether a new design is in the same language.

In 1976 Stiny distinguished between original and analytical grammars. Original grammars enable new design languages to be created, whereas analytical grammars make it possible to understand existing languages.

A parametric shape grammar is an extension of the basic shape grammar formalism and is used to encode neatly a wider range of formal variations for the same rule. In a parametric shape grammar, each rule consists of a set of several rules. By using parametric rules we can encode varying features of shapes so that a greater variety of shapes can be matched to the left-hand side of the rule and then be transformed by the right-hand side.

In this current research, a new type of grammar, called transformation grammar, has been developed to adapt existing dwellings to new requirements. A transformation grammar needs to be parametric because of the variety in the shapes and dimensions of the rooms found in existing dwellings. To transform "rabo-de-bacalhau" dwellings using shape grammars, we needed a grammar that could identify rooms, walls, and spaces whilst taking several features into account, namely area, length, width, function, and material properties. For instance, a bedroom is represented by a quadrilateral shape that satisfies certain requirements such as minimum area, a certain proportional range between length and width, the need for natural light and ventilation, and the need for a door connecting to a circulation area. In this context, a bedroom could be a 4 x 3 m or 3.5 x 4 m rectangle or a 3.5 x 3.5 m square or even a quadrilateral shape with walls that are not perpendicular to each other.

In order to verify the functional adequacy of the original dwellings and the rehabilitation proposals, it was first necessary to determine the fundamental performance criteria by which housing spaces fulfil functional requirements, and then to find a formalism that could be used to analyze spatial configurations from this perspective. The first task was based on Pedro's work [2000], whereas space syntax was used for the functional analysis of spatial configurations.

Space syntax was conceived by Bill Hillier and Julienne Hanson in the late 1970s as a tool to help architects understand the role of spatial configurations in shaping patterns of human behaviour and to estimate the social effects of their designs. In their theory, space is represented by its parts, which form a network of related components. In our research, space is a dwelling in which rooms and circulation areas are connected components with different permeability characteristics. Using space syntax methodology, space is represented first by maps of convex spaces to describe contiguity, adjacency, and proximity and then by graphs in which spaces become nodes and connections become arcs to describe accessibility and permeability.

Research has been developed that combines shape grammar and space syntax in order to formulate, generate, and evaluate designs [Heitor et al. 2004]. In this present research, we used space syntax to provide an accurate means of describing and evaluating spatial properties and therefore, to "increase the likelihood of generating solutions that closely correspond to the user's requirements" [Heitor et al. 2004: 494].

In this research we use shape grammar as a tool to define the methodology for rehabilitating existing types and space syntax as a tool to evaluate the spatial properties of the existing and proposed dwelling designs.

To analyze both the original dwellings and the rehabilitation proposals we considered integration, relative asymmetry, and depth. Integration expresses the degree of space centrality and measures the complexity of reaching this space within a given spatial system. The integration core is obtained by measuring the relative distance of this space from all others in the system. Relative asymmetry (RA) measures the integration of a space by assigning a value between (or equal to) 0 and 1, in which a low value describes high integration. For instance, the degree of integration is usually higher in rehabilitated dwellings than in original ones (see Table 1 below). Depth is a configuration property of a spatial layout. A space is at depth 1 from another if it is directly accessible from it, at depth 2 if it is necessary to go through an intermediate space, and so on [Bellal 2004].

Depending on the strategy followed in the rehabilitation process (see strategies in §3.1), the integration of different functional areas in the dwelling may change. This analysis can be used to choose an adequate rehabilitation strategy, taking family needs and demands into account.

In the current stage of our research, space syntax is not being used to its full potential. We are using graphs to analyze integration, relative asymmetry, depth and control value. In the future, we intend to use the values of these features to evaluate each final rehabilitated design.

2.1.1 Transformable elements: points, lines, surfaces or volumes? A building, a dwelling, or any other kind of construction is defined by its solid mass (e.g., walls, floors, ceilings) and spatial voids (e.g., spaces and rooms). Conventional representations of architecture use lines to represent the boundaries between solid masses and spatial voids, in 2D drawings (figs. 2a, b) or 3D models (fig. 3a). These representations are abstractions of the real objects.

Another, more abstract way of representing architectural space when the exact shape is not of ultimate importance is by means of a graph (fig. 2d,e and fig. 3c). In a graph, the spatial void is represented by a node that connects to other nodes by vectors that are called arcs and represent spatial connections (doors and windows.) In a graph, a spatial void can be represented as a convex spatial void which means that if a room has n different convex spaces, it will be represented with n different nodes.

Different ways of representing dwellings and the transformation rules of the proposed rehabilitation methodology were considered for the current research. The decision regarding which ones to use depended on the architectural elements manipulated in the transformations. The essential elements are walls, doors, and ceilings. Space use (function) was a fundamental attribute of void shapes. In other words, regardless of the architectural representation chosen for the material elements, they had an underlying functional meaning, shown in fig. 2c, d and e, which also had to be included in the representation.

In this context, there were three possibilities:

a. To define rules with justified graphs in which nodes have a functional meaning. This is the most abstract way of representing reality and it limits the possibility of understanding adjacencies, which is extremely relevant in rehabilitation processes. However, at this level of abstraction it is possible to clearly identify the position of a space in the house and characterize its presence, for instance in terms of integration and accessibility (fig. 2e);

b. To define rules with lines and surfaces that represent spatial voids and their functional meaning. This is an abstract way of representing reality and forces us to use the entire spatial void as an element (fig. 2c);

c. To define rules with lines representing the boundaries between solid masses and spatial voids (walls and openings, i.e., doors and windows) in 2D drawings. Using this representation allowed us to: i) assign functional meanings to elements; ii) focus on particular elements (walls and openings) abstracted from the shape of the spaces; and iii) focus on adjacency relationships between different spaces (fig. 2a).

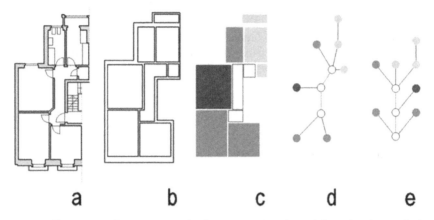

Fig. 2. Different ways of representing a dwelling in 2D: a) traditional floor plan; b) set of solid masses; c) set of spatial voids; d) graph or convex map; and e) justified graph. These representation use points, lines, and surfaces

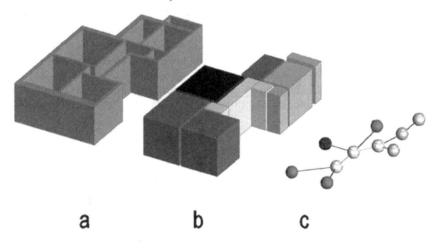

Fig. 3. Different ways of representing a dwelling in 3D: a) set of spatial voids; b) set of solid masses; c) graph

As its advantages suited the goals of the intended transformation grammar better, c) was the representation chosen. It was used to represent the twenty-five designs that constitute the case study and to develop the transformation grammar. In addition, justified graphs were used to complement the description and evaluation of spatial properties.

2.2 Methodology for inferring the transformation grammar: experiments

The methodology used to infer the transformation grammar was divided into three steps, each corresponding to a particular type of experiment:

Step 1: testing the feasibility of the experimental setup by the first author of this present paper and defining a set of preliminary rehabilitation rules that could be transmitted to the experimental subjects in step 2. This step has already been completed (§2.2.1);

Step 2: finding rehabilitation solutions that could satisfy the functional and constructional requirements of each family in a given dwelling. These solutions, designed manually, were used to infer transformation rules. This step has already been completed (§2.2.2);

Step 3: testing the transformation rules inferred in the previous step to confirm whether the solutions generated following these rules were satisfactory. This step has not yet been completed (§2.2.3).

This paper concentrates on the findings of the second step.

The goal was to relate domestic groups (families) to dwellings (existing houses). Prior to applying the methodology, data concerning the domestic groups, case study dwellings, new housing functions, and the pack of ICAT functions was gathered and organized as follows:

1. Families/future inhabitants: five families, differently composed in term of numbers, age, and family relationships, were used in the experiments, namely couples with children, young couples without children, old couples, and couples with children from previous marriages. Initially, people living alone were not considered because the sample dwellings were too big for this family type. Following this, the study focused on how couples with no children or one-person households could be accommodated by dividing each dwelling into two autonomous smaller dwellings, since this accounts for one of the most sought-after types of accommodation in the rehabilitation market in Lisbon [Caria 2004: 163].

2. Existing houses: the housing sample is composed of twenty-five dwellings, some of which were chosen for use in the experiments in the first and second steps. The selection criterion was to choose ten dwellings of varying types (see types, fig. 7) and areas that could potentially satisfy the requirements in the functional programmes for the selected families. To verify this criterion, the area of each dwelling was compared with the area requirements of each family. Two different dwellings that satisfied the criterion were then assigned to each family to obtain ten different dwelling proposals at the end of the experiment (five families x two dwellings.)

3. Functional programme: the minimum functional programme for each family was determined in accordance with Pedro's guidelines and then combined with the requirements expressed by each family in an interview especially designed for this purpose. In the interviews, the families were asked to describe the dwelling they thought they needed, i.e., not an ideal dwelling but one that could fulfil their real needs. They also were told that the dwelling would be in a rehabilitated building with small rooms (the average area of the habitable rooms is 13 m^2) and that they would have to consider economic constraints. The description had to include the

required housing functions or rooms, as well as the topological relations between them. Finally, they were asked to rank their requirements in order of priority.

4. Pack of ICAT functions: in the experiments that were carried out the ICAT pack was reduced to technologies that had an impact on spatial organization (e.g., home cinema, telework, definition of night and day areas for sector alarms). The resulting requirements were then added to the functional programme.

These elements were then given to the experimental subjects in steps 1, 2 and 3.

2.2.1 Step 1. The first and second steps aim to identify the fundamental functional, spatial, and constructional transformations to be carried out on the dwellings studied. The first step consisted of an experiment in adapting the dwellings performed by the first author of this present paper in order to infer some basic transformation rules and to test the feasibility of the exercise, before assigning it to other subjects in steps 2 and 3.

This step included two tasks. The first task consisted of proposing transformations to the dwellings taking the future dwellers' requirements and constructive constraints into account. This resulted in twenty different layout proposals, two for each family/dwelling pair, in order to explore the various possible solutions. The second task consisted of inferring transformation rules from the transformations proposed in the first task. Only higher level transformation rules were inferred, meaning that detailing rules were not considered.

2.2.2 Step 2. The experimental subjects in step 2 were architects with experience in designing houses. The goal of this experiment was to enlarge the set of rehabilitation solutions in order to understand how different approaches may be used to solve the same problem and, therefore, to obtain a larger basis for inferring rules.

This experiment aimed to identify the functional, spatial, and constructive transformations performed manually by human designers, in order to infer the corresponding transformation rules and encode them into a transformation grammar.

In this experiment, the same data from experiment 1 was used, namely, ten existing dwellings and five different families. Two of the architects participating in the experiment were asked to design a solution for all ten family/dwelling pairs (two dwellings for each family,) which yielded twenty different drawn proposals at the end. Three architects designed for five family/dwelling pairs (one dwelling only per family), producing fifteen different drawn proposals.

The experimental tasks were explained to each of the experimental subjects separately and they then completed the work in their offices.

The experimental subjects were asked to perform two tasks: first, using paper or CAD software, to draw a design solution for each family/dwelling pair, taking the functional programme into account as well as the construction constraints; second, to explain the strategy they used to obtain each design proposal.

The data that resulted from these experiments included sketches (two of the architects designed by computer and so did not produce sketches), final drawings of the proposed layouts, and texts explaining the process followed in each case (two of the architects explained the process verbally and so did not write texts.)

The data was analysed and transformation rules are now being inferred. So far, it has been possible to identify two types of design proposals: only one architect proposed transformations like the ones shown in fig. 5b, and four architects proposed transformations like the ones shown in fig. 5c. All the architects respected the given constraints and the priorities expressed by the families. One architect did not comply with the constraint of not demolishing more than 2 meters of wall. In general, they all said that it was difficult to respond to functional requirements because of the original morphology of the dwellings and the demolition constraints.

This step is still in progress and the expected result is a transformation grammar for use in adapting existing dwellings to specific families. §3 describes the present stage of the transformation grammar.

2.2.3 Step 3. The third experiment will be undertaken in a few months' time and will be carried out by two or three architecture students. They will work with dwellings that are part of the corpus but were not used in the previous experiments. The goal is to test the proposed grammar on dwellings that were not used to infer its rules. This will enable us to check whether the inferred rules provide the compositional means for making new transformations in other existing dwellings for other families.

3 Transformation grammar

The definition of a housing rehabilitation methodology is one of the goals of this research. To achieve this goal it is necessary to determine the functional programmes and ICAT packs for specific family profiles. This task can be performed as a standalone process without using the transformation grammar.

However, the use of a specific case study allows us to extend the methodology further. The use of a shape grammar makes it possible to transform existing houses in a very exact and systematic way. Instead of just generating new shapes, as in a traditional shape grammar, a transformation grammar will allow an existing design to be transformed into a new one that matches given requirements, using knowledge that relates family profiles to functional programmes and ICAT packs.

To understand the morphology of the existing building types we carried out a functional, constructional, and social characterization of the buildings, which were constructed between the 1940s and the 1960s. The social characterization was crucial in order to understand the principles that had shaped the existing layouts. Characterization of the construction was necessary in order to define the demolition constraints during the rehabilitation process. Our aim was to make rehabilitation as unintrusive as possible, without compromising comfort and access to ICAT.

The proposed methodology seeks to produce rehabilitated designs that are "legal" because they are in the transformation language and "adequate" because they satisfy the *a priori* set of user requirements [Duarte 2007: 330]. According to Duarte, a grammar applied to an architectural problem must satisfy two functions: it must create or transform an object within a specific language and it must create objects that satisfy the requirements given at the outset. As such, the grammar is structured as a discursive grammar, which includes a shape aspect and a descriptive aspect that evolve in parallel to guarantee that an appropriate dwelling design can be obtained from the description in the functional housing programme. However, unlike Duarte's case, our goal in developing and applying the transformation grammar is not to generate new dwellings using the same language as the existing ones. Our goal is to understand the existing dwellings and the new user requirements prompted by the use of technologies in order to devise transformation principles for adapting existing dwellings to these new requirements.

This grammar clarifies the principles behind the adaptation of the dwellings, such as making circulation more fluid by removing doors in hallways, enlarging social areas by connecting adjacent rooms, and so on. The grammar also encodes principles related to construction constraints, such as avoiding the removal of concrete columns or other structural elements.

The proposed shape grammar is defined in the algebra *U12*, in which lines are combined on a plane. This algebra is augmented by labels in the algebra *V02*, where label points are used to define dwelling functions, and by weights in the algebra *W22*, where shaded surfaces are used to distinguish between the different constructional elements (structural elements, infill brick walls, and light partition walls). The use of shading allows existing infill brick walls that can be taken down to be identified, as illustrated by the rule in fig. 4.

Fig. 4 shows a shape rule that includes a shape part (with shape, labels, and weights – *S, L, W*), a conditional part referring to functional and dimensional aspects, and a descriptive part. Some rules, such as the one shown in fig. 4, have a generic shape that is shared with other rules, whose conditions or descriptions may change.

Rule 1.2.d _ Enlarge the connection between two rooms (by eliminating part of the walls on both sides of the passageway)

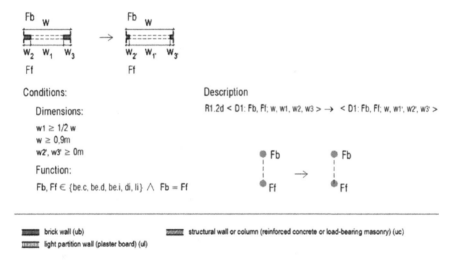

Fig. 4. Example of a rule for enlarging a passageway between two adjacent rooms by partial or total demolition of walls. The shape part is shown at the top and the conditional and descriptive parts are shown at the bottom, on the left and right, respectively

3.1 Three rehabilitation strategies

The process of inferring transformation rules allowed us to identify two possible ways of transforming the dwellings. In addition, a third way of transforming the dwellings was explored in order to create smaller dwellings for households consisting of only one or two people.

Each of these three rehabilitation strategies has advantages and disadvantages in terms of functional and constructional aspects, which can be combined in the same building to generate a wider market offer.

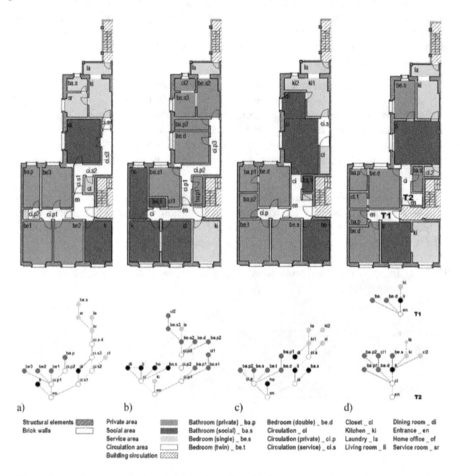

Fig. 5. Rehabilitation strategies: a) original dwelling; b) first strategy; c) second strategy; and d) third strategy. The top row shows the plans and the bottom row the corresponding justified graphs

The buildings in the case study have six to nine floors with a left-right symmetrical layout and two dwellings on each floor. This arrangement is repeated on all floors. The layout of an original dwelling used in the experiments in step 2 is shown in fig. 5a. The results of applying the three rehabilitation strategies to this layout are shown in figs. 5b to d. The differences in the resulting transformations lie in the number of dwellings on each floor and the position of the kitchen in each dwelling.

The three strategies are as follows:

1. Maintain two dwellings on each floor and move the kitchen from its original position in the rear wing of the building (fig. 5b).
   The aim is to strengthen the relationship between social and service areas and to segregate the private area from the rest of the dwelling. The strategy proceeds by:

a) converting the smallest room in the front of the dwelling into the new kitchen, and assigning social spaces (living and dining rooms) to adjacent rooms;

b) occupying the rear wing with private areas (bedrooms and private bathrooms).

2. Maintain two dwellings on each floor and the position of the kitchen (fig. 5c).

The aim is to keep construction transformations to a minimum without compromising the user requirements established in the functional programme. This strategy can be used to rehabilitate just one dwelling in the entire building and is described in more detail in §3.2.2.

3. Divide one dwelling into two smaller ones and create a kitchen in one of the new dwellings (fig. 5d).

The aim is to obtain smaller dwellings and a variety of dwelling types within the building.

a) In the rear dwelling, the social area will preferably be adjacent to the existing kitchen;

b) In the front dwelling, the new kitchen will be located in the smallest room at the front of the building (as in the first strategy) and the social area will preferably be adjacent to the kitchen.

It is possible to combine all three strategies in the same building. Strategies 1 and 3 have a greater constructional impact and require a new vertical drain pipe, as well as a chimney for the new kitchen. The combination of strategies 1 and 2 would have a major constructional impact and would not generate new dwelling types within the building, thereby leading to a narrower market offer. The combination of strategies 1 and 3 would also have a major construction impact but would generate new dwelling types.

The integration (I), relative asymmetry (RA) and depth (TDN) values of the layouts in fig. 5 are shown in Table 1. By comparing the original dwelling with the rehabilitated ones the following may be concluded:

– the first strategy leads to layouts in which the service area is better integrated and has less depth. The circulation and private areas will also be better integrated and have less depth with this strategy;

– the second strategy leads to layouts in which all the areas are better integrated and have less depth.

As the third strategy generates considerably smaller dwellings, the comparison between their syntactic properties and those of the dwellings produced by strategies 1 and 2 was not considered.

In the case presented in this article, the syntactic properties differ according to the rehabilitation strategy chosen. In addition, other cases studied show that such properties also differ according to the initial dwelling types. Such results suggest that the *rabo-de-bacalhau* dwelling types A to D (fig. 7) have particular characteristics which enable specific rehabilitation strategies to be carried out. This observation leads us to believe that the combination of rehabilitation strategies and dwelling types allows us to design for various lifestyles and thus meet current market demands.

	Original dwelling			First strategy			Second strategy			Third strategy (T1/T2)		
	i	RA	TDN	I	RA	TDN	i	RA	TDN	i	RA	TDN
Min	1.6	0.2	50	1.9	0.2	46	2	0.2	40	1.2/1.9	0.2/0.2	5/17
Mean	2.8	0.4	70	3.2	0.3	68.6	3.1	0.3	57.2	2.6/3.3	0.5/0.4	7.2/25.8
Max	4.1	0.6	104	5.5	0.5	97	5	0.5	76	6/6.4	0.8/0.5	9/34
Private area	2.26	0.44	77.5	2.86	0.36	73.2	2.57	0.39	63.2	1.5/2.67	0.66/0.39	8/27.6
Social area	3.17	0.32	61	2.59	0.38	77	3.46	0.29	51.5	3/4.22	0.33/0.26	6/22
Service area	2.07	0.49	84.8	2.73	0.37	75	2.22	0.46	71	1.2/2.5	0.83/0.41	9/29.5
Circulation area	3.43	0.29	57.7	4.58	0.22	52	4	0.26	47.4	6/4.62	0.16/0.25	5/21.5

Table 1. Integration (i), relative asymmetry values (RA) and total depth for current node (TDN) of original dwelling and the three rehabilitation strategies (all calculations performed using AGRAPH software)

Fig. 6 shows the simplified derivation and decision tree for the three rehabilitation strategies described above, which is reflected in the structure of the grammar described in the following section.

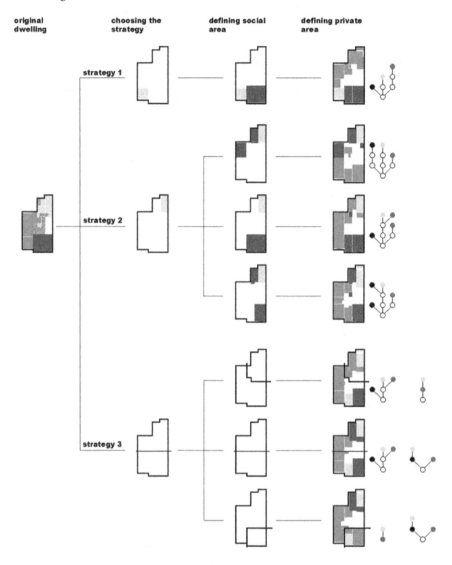

Fig. 6. Simplified derivation tree for the different rehabilitations strategies

3.2 Rules of the Transformation Grammar

The different experiments undertaken during the process of inferring the rules of the transformation grammar revealed certain rehabilitation patterns. The decision-making processes used by the experimental subjects tended to be similar and to follow the same sequence for major decisions, for instance, the location of private and social areas.

As described in the previous section, there are three different rehabilitation strategies for the buildings in the case study, but in developing the grammar we decided to consider the one with the least constructive impact first, namely the second strategy, shown in fig. 5c. The three rehabilitation strategies are briefly described in the next section. The rules corresponding to the second strategy are described in §3.3.3 and were inferred from experiments in which the solutions generated followed this strategy.

3.2.1 Choosing an appropriate dwelling. The first major decision in the rehabilitation process is to choose an adequate dwelling for a given family. There are three possible market scenarios in which such a decision will have to be made: a family looking for a dwelling to rehabilitate, a family intending to rehabilitate its current dwelling, and a property developer intending to rehabilitate an existing dwelling for sale or for rent, in which case the future dwellers are unknown.

In any of these cases:

- if a family is seeking to buy a dwelling to rehabilitate, it would be necessary to identify a set of dwellings of the type that could accommodate the family's functional programme;
- if a family intends to rehabilitate its current dwelling, it would be necessary to assess whether this dwelling would respond to the family's functional programme and, if not, to propose another dwelling (fig. 8);
- if a property developer intends to rehabilitate an existing dwelling for sale or for rent, it would be necessary to carry out market analysis to estimate the potential buyer or tenant family type. In this case, the future dwellers are anonymous.

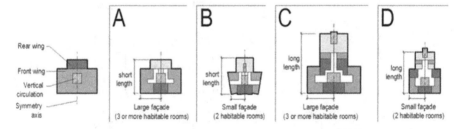

Fig. 7. Types of *rabo-de-bacalhau* dwellings

The selection of an appropriate dwelling or the assessment of the appropriateness of an existing one is based on two criteria: i) the correspondence between the type of dwelling (fig. 7) and the family structure, and ii) the correspondence between the area needed to accommodate the functional programme and the area of the dwelling. These data are summarized in data tables not shown in this paper. The area needed to accommodate the functional programme is an essential parameter that allows us to select a dwelling type from among the ones included in the case study, labelled A to D. The data tables show that types A and C have larger areas and that types B and D are more appropriate for smaller families. Despite this, all the types except type B can be subdivided using the third strategy (fig. 5d).

When looking for an appropriate dwelling the following two scenarios are possible:

a. looking for an original dwelling;
b. looking for a dwelling that has been divided into two smaller ones, in which case the proposed dwelling has no, one, or two bedrooms, i.e., it is T0, T1 or T2 accommodation, respectively.

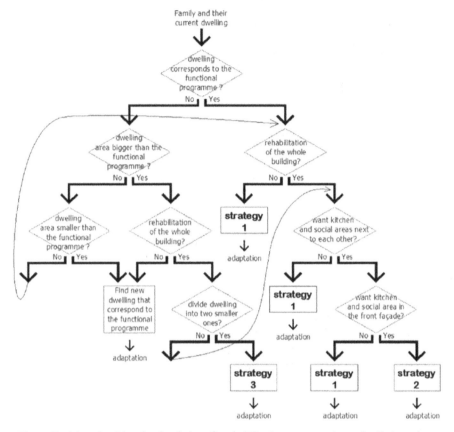

Fig. 8. Decision algorithm for the choice of a rehabilitation strategy when a family intends to rehabilitate their dwelling

3.2.2 Adaptation of the dwelling. The next decision concerns the adaptation of the family's dwelling to the new functional programme and the required ICAT pack. As mentioned above, we will show how adaptation proceeds according to the second rehabilitation strategy. This strategy aims to have the least construction impact without compromising the requirements established by the functional programme and the ICAT pack. It can be applied to just one dwelling or to an entire building. Following the experiment results, the list of steps in this strategy has been systematized as shown in Table 2 and is explained below. Steps 0 to 6 are currently being developed. Step 7 will be developed in the coming months. Due to space limitations, in this paper we only show a sample of the transformation rules in order to illustrate how the grammar works and can be applied.

All the transformations proposed in the grammar involve the construction or demolition of walls, as well as the assignment of functions to rooms. Therefore, the proposed rules include the following types of actions: i) adding walls, which enables rooms to be divided and wall openings to be removed or reduced; ii) eliminating walls, which enables adjacent rooms to be joined or rooms to be connected together; iii) the assignment of functions to spaces; and iv) changing functions between spaces.

0	Assignment of kitchen / according to the chosen strategy		
1	If functional programme has 2 or more bedrooms → defining private area	1.1	***Assignment of bedrooms***
		1.2	***Adapting bedroom shapes***
		1.3	Assignment of private bathroom(s)
		1.4	Adapting private bathroom shape(s)
2	If functional programme has 2 or more bedrooms → defining social area	2.1	Assignment of living room, dining room, home office, media room
		2.2	Adapting living room, dining room, home office, media room
		2.3	Assignment of social bathroom
		2.4	Adapting social bathroom shape
3	Defining circulation (social and private area)	3.1	Assignment of circulation
		3.2	Adapting circulation
4	Defining service area	4.1	Adapting kitchen shape
		4.2	Assignment of laundry room
		4.3	Adapting laundry room shape
		4.4	Solving circulation in the service area
5	Defining storage spaces	5.1	Assignment of clothes storage
		5.2	Assignment of general storage
		5.3	Solving circulation involving storage spaces
6	Solving circulation between different circulation spaces		
7	Incorporation of ICT elements (This part of the shape grammar will be developed in the coming months; only some examples of steps are presented here)	7.1	Introduction of water detectors
		7.2	Introduction of presence detectors
		7.3	Introduction of temperature sensors
		7.4	Introduction of smoke detectors
		7.5	Introduction of control panels
		7.6	...

Table 2. Steps to follow in adapting a dwelling. Steps 1.1 and 1.2 (in bold) are described in greater detail in the text

Step 0: Assignment of kitchen

The grammar starts by locating the kitchen. In accordance with the chosen strategy, the kitchen can be assigned to one of the following locations:

- First strategy: move the kitchen from its original position to the front of the building. Call rule 0.1c or rule 0.1d;

- Second strategy: maintain the kitchen position. Call rule 0.1a or rule 0.1b;

- Third strategy: maintain the kitchen position at the rear of the dwelling and assign a new kitchen to the front of the dwelling.

Step 1: Defining private area

The grammar continues by locating the private and social areas, proceeding as follows:
   a. If a private area requires two or more bedrooms, this area should be located first (step number 1), either at the front or rear of the building. Call rule 1.1;
   b. If a private area requires fewer than two bedrooms, the social area should be located first (step number 2), at the front or rear of the building. Call rule 2.1.

Rule 1.1: Assignment of bedrooms

With this rule we can assign a bedroom function to an existing space (fig. 9). The left-hand side of the rule verifies whether an existing space has the features required for it to become a bedroom. Specifically, considering the required number of bedrooms n defined in the functional programme, it determines n rooms that satisfy the following requirements:

   – Bedrooms are rooms with natural light and ventilation;
   – Bedrooms have an area equal or superior to 7 m^2 (single bedroom), 9 m^2 (twin bedroom) or 10.5 m^2 (double bedroom);
   – Bedrooms must have direct access to a circulation area;
   – Connections between two bedrooms must be via a private circulation area or, in exceptional cases, through a service circulation area;
   – In dwelling types A or B, bedrooms will be located at the front of the dwelling (second strategy);
   – In dwelling types C or D, bedrooms may be located at the front or rear of the dwelling.

Some of these requirements can be overridden by specific family requirements. For instance, a family might want a bedroom that can be accessed directly from the living-room.

Regra 1.1.a assignment of bedrooms (minimum level)

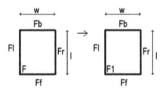

Conditions:

Dimensions:

$7\text{m}^2 \leq F < 9\text{m}^2 \wedge w, l \geq 2,1\text{m} \Rightarrow F1 \in \{\text{be.s}\}$

$F < 10,5\text{m}^2 \wedge w, l \geq 2,1\text{m} \Rightarrow F1 \in \{\text{be.t, be.s}\}$

$F \geq 10,5\text{m}^2 \wedge w, l \geq 2,7\text{m} \Rightarrow F1 \in \{\text{be.d, be.t, be.s}\}$

Functions:

$Fb \vee Ff \vee Fr \vee Fl \in \{\text{ext}\} \wedge Fb \vee Ff \vee Fr \vee Fl \in \{\text{ci, cl, dr}\}$

Description:

Assignment of single bedroom

R1.1a < D1: ext, {ci, cl, dr}, Ff, Fl; F; Z > →

   < D1: ext, {ci, cl, dr}, Ff, Fl; be.s ; Z - {be.s} >

Assignment of twin bedroom

R1.1a < D1: ext, {ci, cl, dr}, Ff, Fl; F; Z > →

   < D1: ext, {ci, cl, dr}, Ff, Fl; be.t ; Z - {be.t} >

Assignment of double bedroom

R1.1a < D1: ext, {ci, cl, dr}, Ff, Fl; F; Z > →

   < D1: ext, {ci, cl, dr}, Ff, Fl; be.d ; Z - {be.d} >

Fig. 9. Rule 1.1 Assignment of bedrooms (minimum level). The shape part is shown at the top and the conditional and descriptive parts are shown at the bottom, on the left and right, respectively

Rules 1.2: Adapting bedroom shapes

To clarify and reduce the number of rules presented here we classify the rules for adapting bedroom shapes in the following four spatial configurations, represented by means of a graph (fig. 10).

Adding or appending   Closing or widening the   Changing door position   Reducing bedroom area
an adjacent space   connection with an
            adjacent space

Fig. 10. Possible topological transformations involved in the adaptation of bedroom shapes. Dark gray nodes represent bedrooms and white nodes represent other spaces. Lines between nodes represent connections between spaces. Dashed lines represent widened connections.

The use of graphs helped us to group rules according to the topological transformations involved, without considering specific shapes. For instance, several rules may be used to add an adjacent space, but in all cases two nodes are merged into one. Therefore, graphs are used in a parallel grammar where nodes and arcs are descriptions of space configurations. From among these transformations of space configurations, we will only exemplify the ones that manipulate a connection between two adjacent spaces (fig. 11).

In fig. 12 we show a possible derivation of the rehabilitation proposed in Figure 5c following the second strategy. The final layout of this rehabilitation is a result of the experiments carried out to infer the grammar (see §2.2.2, Step 2) and was designed for a specific family with a specific functional programme.

4 Conclusions

Information plays an increasingly important role in our lives and, as a consequence, ICAT is changing the ways in which we inhabit spaces. Recent intelligent technologies aim to maximise the use of information with the dual goal of providing houses with increased access to information and comfort through the use of automated control systems.

In addition, the rehabilitation of Lisbon's existing housing stock in order to meet the new space-use requirements demanded by new information age lifestyles is a priority in the political agenda.

Our ongoing research aims to identify strategies that allow an adequate balance to be achieved between these goals. This paper explores the use of shape grammar formalism in a rehabilitation context.

In this context, formalism is used to encode the rules for transforming existing dwellings into new ones adapted to contemporary lifestyles. In addition, space syntax is used to form a parallel description grammar to guarantee that designs with adequate functional organization are generated. The resulting compound grammar is proposed as a way of developing and encoding a methodology for housing rehabilitation that can easily be explained to, and applied by, architects.

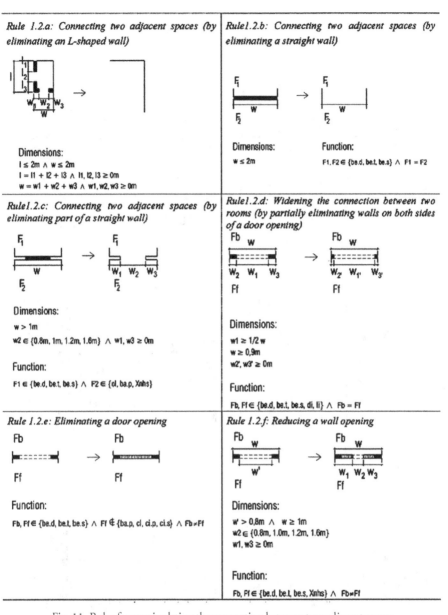

Rule 1.2.a: Connecting two adjacent spaces (by eliminating an L-shaped wall)

Dimensions:
l ≤ 2m ∧ w ≤ 2m
l = l1 + l2 + l3 ∧ l1, l2, l3 ≥ 0m
w = w1 + w2 + w3 ∧ w1, w2, w3 ≥ 0m

Rule1.2.b: Connecting two adjacent spaces (by eliminating a straight wall)

Dimensions:
w ≤ 2m

Function:
F1, F2 ∈ {be.d, be.t, be.s} ∧ F1 = F2

Rule1.2.c: Connecting two adjacent spaces (by eliminating part of a straight wall)

Dimensions:
w > 1m
w2 ∈ {0.8m, 1m, 1.2m, 1.6m} ∧ w1, w3 ≥ 0m

Function:
F1 ∈ {be.d, be.t, be.s} ∧ F2 ∈ {cl, ba.p, Xnhs}

Rule1.2.d: Widening the connection between two rooms (by partially eliminating walls on both sides of a door opening)

Dimensions:
w1 ≥ 1/2 w
w ≥ 0,9m
w2', w3' ≥ 0m

Function:
Fb, Ff ∈ {be.d, be.t, be.s, di, li} ∧ Fb = Ff

Rule 1.2.e: Eliminating a door opening

Function:
Fb, Ff ∈ {be.d, be.t, be.s} ∧ Ff ∉ {ba.p, cl, ci.p, ci.s} ∧ Fb≠Ff

Rule 1.2.f: Reducing a wall opening

Dimensions:
w > 0,8m ∧ w ≥ 1m
w2 ∈ {0.8m, 1.0m, 1.2m, 1.6m}
w1, w3 ≥ 0m

Function:
Fb, Ff ∈ {be.d, be.t, be.s, Xnhs} ∧ Fb≠Ff

Fig. 11. Rules for manipulating the connection between two adjacent spaces

Fig. 12. Partial derivation of one rehabilitation transformation

The transformation rules of the grammar were inferred following a set of experiments in which several architects were asked to rehabilitate specific dwellings for specific families. The rehabilitated dwellings belong to a building type used as a case study because it accounts for a significant number of buildings in Lisbon. The experiments revealed patterns of design decisions and transformations which made the development of the grammar feasible.

Future work will be concerned with completing the compound grammar and fine-tuning the relationship between the shape grammar rules and the space syntax graph descriptions.

Acknowledgments

The research described in this paper was funded by the Portuguese Science Foundation (FCT) through grant SFRH / BD / 18225 / 2004. We thank I. Plácido and R. Nunes for their contributions to this research.

References

BELLAL, Tahar. 2004. Understanding home cultures through syntactic analysis: the case of Berber housing. *Housing, Theory & Society* **21**, 3: 111-127.

CARIA, Helena (coord.). 2004. *Habitação e Mercado Imobiliário na Área Metropolitana de Lisboa.* Lisboa: CML, Pelouro de Licenciamento Urbanístico e Planeamento Urbano.

DUARTE José Pinto. 2001. Customizing Mass Housing: A Discursive Grammar for Siza's Malagueira houses. Thesis submitted to the Department of Architecture in partial fulfilment of the requirements for the Degree of Doctor of Philosophy in Design and Computation. Massachusetts Institute of Technology, Cambridge, E.U.A.

————. 2007. *Personalizar a habitação em série: Uma gramática discursiva para as casas da Malagueira do Siza.* Lisboa: Fundação Calouste Gulbenkian, Fundação para a Ciência e a Tecnologia.

HEITOR, Teresa V., José P. DUARTE, and M. RAFAELA. 2004. Combining Grammars and Space Syntax: Formulating, Generating and Evaluating Designs. *International Journal of Architectural Computing* **2**, 4 (Dec. 2004): 492-515 (24).

KNIGHT, Terry W. 1994. *Transformations in design. A formal approach to stylistic change and innovation in the visual arts.* Cambridge University Press.

————. 2000. *Shape Grammars in education and practice: history and prospects.* http://web.mit.edu/tknight/www/IJDC/ (last accessed 22 November 2010).

MANUM, Bendik, Espen RUSTEN and Paul BENZE. 2005. AGRAPH, Software for Drawing and Calculating Space Syntax Graphs. *Proceedings of the 5th International Space Syntax Symposium*, Delft 13-17 June 2005. http://www.spacesyntax.tudelft.nl/media/Long%20papers%20I/agraph.pdf. (Last accessed 15 November 2010)

PEDRO, João Branco. 2000. Definição e avaliação da qualidade arquitectónica residencial. Ph.D. thesis. Lisbon: Ed. Faculdade de Arquitectura da Universidade do Porto.

STINY, George. 1976. Two exercises in formal composition. *Environment and Planning B: Planning and Design* **3**, 2: 187-210.

————. 1980. Introduction to shape and shape grammars. *Environment and Planning B: Planning and Design* **7**, 3: 343–351.

————. 1992. Weights. *Environment and Planning B: Planning and Design* 19, 4: 413-430.

STINY, George and James GIPS. 1972. Shape Grammars and the Generative Specification of Painting and Sculpture. Pp. 1460-1465 in *Information Processing 71*, C. V. Freiman, ed., Amsterdam: North Holland.

STINY, George and Lionel MARCH. 1981. Design machines. *Environment and Planning B: Planning and Design* **8**, 3: 245-255.

About the authors

Sara Eloy graduated in Architecture (1998) and is currently a Ph.D. Student at the Instituto Superior Técnico, Universidade Técnica de Lisboa and Laboratório Nacional de Engenharia Civil. The starting point for her Ph.D. thesis is the premise that the future of the real estate market in Portugal requires the rehabilitation of existing residential areas. This should envisage the incorporation of Information and Communications Technologies (ICT) and domotics, as well as responding to the new demands of dwellers. Sara is a Lecturer in the Department of Architecture and Urbanism at ISCTE-IUL in Lisbon, where she teaches courses on building technology and computation. Her research interests include ICT, home automation, rehabilitation, shape grammar, green architecture, building technologies, building accessibility, and space syntax.

José Pinto Duarte holds a B.Arch. (1987) in architecture from the Technical University of Lisbon and an S.M.Arch.S. (1993) and a Ph.D. (2001) in Design and Computation from MIT. He is currently Visiting Scientist at MIT, Associate Professor at the Technical University of Lisbon Faculty of Architecture, and a researcher at the Instituto Superior Técnico, where he founded the ISTAR Labs - IST Architecture Research Laboratories. He is the co-author of *Collaborative Design and Learning* (with J. Bento, M. Heitor and W. J. Mitchell, Praeger 2004), and *Personalizar a Habitação em Série: Uma Gramática Discursiva para as Casas da Malagueira* (Fundação Calouste Gulbenkian, 2007). He was awarded the Santander/TU Lisbon Prize for Outstanding Research in Architecture by the Technical University of Lisbon in 2008. His main research interests are mass customization with a special focus on housing, and the application of new technologies to architecture and urban design in general.

José Nuno Beirão

Faculty of Architecture
Delft University of Technology
Julianalaan 134
2628 BL Delft
THE NETHERLANDS
J.N.Beirao@tudelft.nl

José Pinto Duarte*
*Corresponding author

Faculdade de Arquitectura
Universidade Técnica de Lisboa
Rua Sá Nogueira
Pólo Universitário, Alto da Ajuda
1349-055 Lisbon PORTUGAL
jduarte@fa.utl.pt

Rudi Stouffs

Faculty of Architecture
Delft University of Technology
Julianalaan 134
2628 BL Delft
THE NETHERLANDS
r.m.f.stouffs@tudelft.nl

Keywords: Urban design, shape
grammars, urban patterns

Research

Creating Specific Grammars with Generic Grammars: Towards Flexible Urban Design

Presented at Nexus 2010: Relationships Between Architecture and Mathematics, Porto, 13-15 June 2010.

Abstract. The aim of the City Induction project is to develop an urban design tool consisting of 3 parts: an urban programme formulation module, a generation module and an evaluation module. The generation module relies on a very generic Urban Grammar composed of several generic grammars called Urban Induction Patterns (UIPs) corresponding to typical urban design moves. Specific grammars, such as the analytical grammars inferred from our case studies, can be obtained by defining specific arrangements of Urban Induction Patterns and specific constraints on the rule parameters. We show that variations on the UIP arrangements or rule parameters can provide design variations and specific grammars to be synthesised through design exploration. It is therefore seen as a process for synthesizing a specific design grammar within the field of urban design and has two main features: (1) it allows for the synthesis of specific grammars during the design process and (2) it allows for the customization of a personal design language within the broad scope of the generic grammar.

A formal definition of Urban Grammars is presented and its application in the production of customized urban designs is demonstrated by customizing design languages using a specific compound grammar defined by a specific arrangement of generic grammars.

Introduction

1 The City Induction Project

This paper presents a detailed description of the generation module for the City Induction project. The City Induction project aims to develop an urban design tool which is defined by linking three operative modules through a common ontology, integrating knowledge structures and representations of cities.

The three modules are:

- a formulation module, which reads data on a site's context and formulates urban programme specifications for that site (produces the urban programme) [Montenegro and Duarte 2008];
- a generation module, which generates alternative urban design solutions for the same site (produces design solutions);
- an evaluation module, which guides the generation to meet the programme's goals (guides the generation towards satisfactory designs) [Gil and Duarte 2008].

Nexus Network Journal 13 (2011) 73–111
DOI 10.1007/s00004-011-0059-3; *published online* 25 February 2011
© 2011 Kim Williams Books, Turin

The formulation module provides its output by means of goal patterns encoded into a description grammar [Stiny 1981]. The generation module uses arrangements of generic discursive grammars [Duarte 2001] to generate urban designs, and the evaluation module uses several evaluation techniques to guide generation towards the specified requirements. In this paper we show how a set of generic discursive grammars can be used to design specific grammars for urban design.

2 Defining the problem

Uncertainty and complexity seem to be dominant paradigms in the growth of cities. The main problem is that, even when planned, the development of cities is difficult to predict. Designing cities involves the ability to deal with many simultaneous and complex city development behaviours and all their components, and predict desirable and controllable city developments. This has been proved virtually impossible to achieve, at least by traditional means [Portugali 1999]. In addition, the constantly changing city dynamics in contemporary society has led to the growing inefficiency of the traditional layout planning approach. Flexibility and adaptability have become imperative [Archer 2001].

In order to progress towards more efficient design systems we need to develop very flexible and interactive platforms that are able to assess the complexity of urban systems without interfering with the typical indeterminate design exploration procedures that designers adopt. Previous work [Beirão and Duarte 2009] has shown that designing urban plans with shape grammars [Stiny and Gips 1972] establishes planning systems containing explicit and implicit flexibility that can be used as adaptive features in a real-world implementation where such features become extremely important. The idea is that it is possible to define design systems that establish an embedded order through a set of design rules whilst still retaining the adaptive features that can accommodate uncertainties. In practical terms, the aim is to develop an urban design platform that can shift from the rigid layout paradigm to a new concept, the concept of city information modelling (CIM), and eventually extend the term modelling to monitoring by incorporating the analysis and evaluation tools provided by the evaluation module.

However, the implementation of shape grammars contains problems of its own. The problem of shape recognition [Yue et al 2009] has been pointed out many times as its main technical restriction, whilst it is claimed that the mathematical definitions, being founded on visual reasoning, support the type of visual ambiguity found in design [Knight 2003; Stiny 2005]. Moreover, ambiguity conflicts with design control, whilst the definition of a very detailed and complete grammar conflicts with design freedom. The latter problem arises from the fact that a shape grammar, even if parametric, always embeds some kind of design language, imposing the inherent language on the designs generated. The problem that concerns us and led to this research is that design languages are the result of design synthesis and not the reverse, meaning that exploring design languages is not an aim of design, whereas the synthesis of a design language is. Exploration of the language is only an extension of the design capacities, not the purpose of design itself. As such, the main question therefore is how to define a shape grammar during the exploratory design process.

Therefore, the research questions are, firstly, how to design using shape grammars, given that a designer's language is usually synthesized together with the design process itself and, secondly, how to apply shape grammars in urban design in order to obtain a more flexible and efficient urban design process. Solving the first problem provides a response to the second, as it forms the basis for the development of a supporting design tool. However, the second question involves complex features of the urban environment that make it much more difficult to answer. This research falls within the framework of urban design.

To address these problems the research used the following methodology. First, the main characteristics of the existing urban design and assessment tools were assessed, in order to figure out how shape grammars should be used in conjunction with these systems. A survey was also carried out of the supporting literature capable of providing guidance in specifying the aims of the design tool – i.e., what it should do and what it does not need to do. The following section presents the theoretical background to the research. In the third section, we propose the use of arrangements of generic grammars as a means of enabling specific design grammars to be developed. Section 4 shows the technical definitions of this concept, the structure of the generation module and how it works within the City Induction concept. Section 5 contextualises the research design space within the framework of four case studies, in order to simplify the prototype implementation, starting from the assumption that the natural approach in this case is to progress from simple to complex implementations. In this section we present some grammar examples extracted from the case studies to demonstrate that generic grammars correspond to urban induction patterns and that specific arrangements of generic grammars produce specific grammars. We may therefore call generic grammars designing grammars. The discussion section engages the reader in a critical review of the achievements of the research and the scope for future work, establishing some new hypotheses for future research. The final section draws conclusions on the achievements of the research.

3 Background and theoretical support

Architecture, urban design and urban planning are three different scales of design activity that merge within the context of the city. It is already established in literature on the subject that these scales range from local to global (or vice versa) alongside several complex interactive systems, namely social, economic, environmental and political, all of which contribute towards generating uncertainty and complexity in cities and their development [Archer 2001; Batty 2005; Portugali 1999]. In this environment, urban design becomes a rather difficult and unpredictable task.

In the current state of the art, it is impossible to find fully integrated tools that enable us to assess the many aspects relating to the task of urban design. There are several tools for urban analysis, tools for evaluation and tools for designing but no single tool seems adequate to assess all the demands of urban design. The basic distinction that is important to this paper is the distinction between GIS and CAD tools. The former are extremely powerful assessment tools for evaluating urban data and performing many analytical tasks which may inform urban design, but they are not design tools. GIS interfaces share the characteristic of gathering geo-referenced information and representations of existing components or concept-components[1] in our environment. Data and shape-files (i.e., representations) are linked by a geographical reference. GIS platforms provide many different tools that allow us to run several types of analysis. There are also other types of software or plug-ins that add other analytical functions to these platforms, space syntax [Hillier and Hanson 1989; Hillier 1998] being probably one of the most widely-used tools of this kind. Although some of these types of software might contain some editing tools, none of them are drawing or modelling tools and for this reason, GIS platforms are very unfriendly tools for design purposes.

On the other hand, CAD tools are essentially drawing or modelling tools that do not assess data, nor do they allow for the complete topological integration of representations and data. However, most of the CAD software is already very powerful and versatile in terms of design purposes, although communication between the different platforms is difficult and implies loss of data. Nevertheless, it is clear that urban design methodologies are strongly supported by intensive analytical methods during the pre-design phases, therefore indicating the enormous potential and desirability of linking GIS and CAD to allow for analysis-design-analysis data flow cycles. Establishing the foundations of a tool for this purpose is one of the main goals of City Induction. On an urban scale, this is actually a direct translation of what Schön [1983] would call a see-move-see cycle in architectural design.

A shape grammar [Stiny and Gips 1972] is a set of shape transformation rules that are applied recursively to generate a set of designs. These are $\alpha \rightarrow \beta$ type rules in which α and β are labelled shapes from a finite set of shapes S and a finite set of labels L. The rule finds the occurrence of a transformation τ of the labelled shape α in a design δ and replaces it with a transformation τ of the labelled shape β as defined in the equation $\delta' = [\delta - \tau(\alpha)] + \tau(\beta)$, where δ' is the resulting design after the rule iteration and $-$ and $+$ are the Boolean difference and union operations [Stiny 1980]. As Stiny has pointed out, shape recognition is an ambiguous task [2005] and needs correctly supported artificial intelligence to be effective in a computer-based implementation of shape grammars. Finding $\tau(\alpha)$ can prove a very complex task when new shapes emerge during design generation. In addition, extending shape grammars to the space of parametric grammars, which are in fact used in most design situations, makes this even more difficult, as the recognition of a shape becomes the recognition of any assignment g of parameter values to a parametric labelled shape α, i.e., finding $g(\alpha)$ in a design δ to apply the rule schemata $g(\alpha) \rightarrow g(\beta)$.

Shape grammars have successively demonstrated a capacity to encode the design rules embedded in design languages with a rigorous technical formalism. However, the semantic discourse in urban design is not only provided by shape transformations but also political, social and territorial contexts which are informed by features other rather than those of form. To solve this semantic problem previously pointed out by Fleisher [1992], Duarte [2001] proposed the concept of discursive grammars, a combination of description grammars [Stiny 1981] and shape grammars, as a way of providing descriptions of designs that are appropriate for a particular context.

Previous work using shape grammars and patterns [Alexander et al 1977] as an approach to solving urban design problems has been carried out in recent years in design studios at the Technical University of Lisbon [Beirão and Duarte 2009]. This work still remains the main motivation for the City Induction research project, since it has demonstrated the potential of using shape grammars in urban design. Although the design studio was run with current tools, shape grammars were applied informally using current CAD functionalities and the analytical tasks were performed without the support of GIS-based tools, the idea of integrating analysis, generation and evaluation into a single working platform has been our main focus since then.

4 Designing grammars for urban design

As previously mentioned, the implementation of shape grammars implies specific technical problems. Basically, two main problems concerning shape grammars are addressed in this research: firstly, the shape recognition problem and secondly, the problem of defining the grammar during the design process. The linking of GIS and CAD representations into a compatible format defines the third problem under investigation.

Our current work focuses on the implementation of the generation module for City Induction. It proposes two devices as a means of solving the three said problems. The first problem is partially solved by the introduction of a City ontology. The third problem is entirely solved by the same device. The second problem is solved through the introduction of Urban Induction Patterns (UIPs), small generic grammars encoding urban design moves.

The City ontology defines and organizes significant relationships between the various types of objects or components found in the urban space that will be used in the urban design process. It is structured into object classes, each containing object types and attributes. At top level the City ontology contains 5 different classes – networks, blocks, zones, landscape and focal points. The ontology was defined to support communication between the three modules of City Induction, but also provides a structure that can create layered representations of city features. This layered structure is envisaged as a means of establishing the direct export of design generations to a GIS platform. It is also seen as a way of structuring urban grammars into parallel generic grammars, basing the definition of the shape sets of these grammars on the object classes of the ontology. Details on the ontology can be found in a recent paper [Beirão, Montenegro et al 2009].

In this paper we show how design moves [Schön 1983] can be encoded into small generic grammars and combined in different ways to form customized designing grammars that can therefore be used to synthesize a personal design language during the design process. Such a system may be used to improve design procedures and design exploration, as it enables the advantages of the generative properties of shape grammars to be used, whilst also allowing a personal design language to be explored in the design.

Encoding design moves into generic grammars: Urban Induction Patterns (UIP)

Donald Schön says that designs evolve through a series of see-move-see cycles. It is a reflective process that is performed continuously throughout the design process by the designer until s/he comes up with a proposed final solution. The design rules are the result of such a process. Only when the process is considered finished, is the designer able to talk about the design process and replicate the procedure. Only then, is the design capable of providing a specific grammar, a set of shape rules that can translate the design language of the architect, i.e., a consistent design expression translated into a shape grammar.

What this research intends to develop is a way to simultaneously provide the exploratory design process whilst also developing a consistent design grammar encoded into the algebraic formalism of shape grammars, so that by the end of the design process the designer can obtain a complete customized shape grammar that enables him to further explore the design space defined by the grammar. In the case of urban design, what this process provides is the possibility of designing, not a final layout, but a set of rules that can produce any layout within the design space defined by the grammar. A design language is therefore provided without enforcing a specific layout.

The main idea underlying this work is that we can "break down" the complete design process into a particular arrangement of independent design moves. Each design move is encoded into a very generic shape grammar independent of context that can be applied to different contexts and customized by constraining the available parameters.

Generic grammars are very simple, customizable, context-independent shape grammars corresponding to generic design moves. They can be defined because what is encoded is not a complete design sequence but short recurrent design moves common to most designers. Expressions such as "defining the main axis", "placing a landmark" or "setting the grid" are understood by any urban designer, although each individual might have a specific interpretation of their meaning. These design moves are set as design patterns [Gamma et al 1995], defining a short and very generic piece of code that generates this specific design move. Specific designs are the result of composing an arrangement of such design moves and setting specific values for the available parameters. A specific grammar is therefore the result of setting a specific arrangement of generic grammars and constraining the available parameters to the ones that express the designer's thought language. In order to simplify the text which follows, a generic grammar encoding a recurrent urban design move will be called an Urban Induction Pattern or simply UIP.

Generic grammars for designing Urban Grammars

Another important aspect to consider is that design moves are a designer's response to specific stimuli found in the design context. In other words, certain elements found in the context provide referential support for design decisions. These elements or references are the ones that provide the initial shapes to which a generic grammar can be applied. The references are elements that the designer selects in the design context or referential elements generated by previously applied grammars. There are therefore two ways of providing referential elements or initial shapes in order to apply a UIP: firstly, the designer selects elements in the design context and gives them the attribute R_{ef} (for referential element); alternatively, a previous grammar generates an element that is then used by another generic grammar as an initial shape.

The City Induction generation module is a design generation system based on an extremely generic Urban Grammar defined by the available UIPs and rule parameters. A specific Urban Grammar is customized through the selection of a specific arrangement of UIPs and specific constraints on the rule parameters. A specific design is instantiated for a specific selection of references (R_{ef} elements) by instantiating specific values within the constraints defined for the rule parameters. In terms of the design decision process, a specific design is obtained through three kinds of design decisions: selecting or defining references in the design context, selecting a specific arrangement of UIPs, and constraining the available parameters within the pattern rules.

These options depend on the decision taken by the formulation module or the designer. Therefore, the design process is a reflective process, responds to contextual features including regulations and quality standards, and is rule-based.

The various meanings of the term "pattern"

A brief digression is necessary here in order to explain and identify the meaning of the word "pattern" in Urban Induction Patterns. The basis of the word "pattern" as used here was first coined by Alexander et al in the book *A Pattern Language* [1977] and was defined as a recurrent problem occurring in the environment that can be clearly identified and provided with a generic solution. This concept was later extended by Gamma et al [1995] so that it could be applied in object oriented programming. The underlying idea was to identify recurrent problems in object oriented programming and to provide a detailed solution with a sample code for generic application. This extended concept was called "design pattern", although the word "design" here means software design. An Urban Induction Pattern is a compound version of both concepts applied to urban design, i.e., an UIP is a recurrent urban design move provided with a generic grammar that replicates this design move and can be applied in many different contexts. In every sense in this context these are design patterns.

The next sub-section details this definition in terms of grammar formalisms.

Urban Induction Patterns and Urban Grammars: definitions

The top level of the City ontology defines 5 object classes, namely Networks, Blocks, Zones, Landscapes and Focal Points (fig. 1). Each object class is a set of object types divided into two major subsets: geometry (shapes, parameterized shapes) and attributes (labels). The ontology provides a dependency structure for all the shape and label sets that comprise an urban plan, allowing grammars to accept them as parameterized shapes and labels for the application of rules used to generate urban designs.

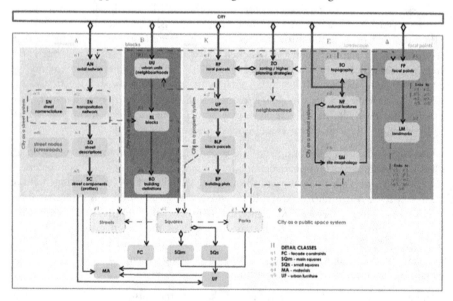

Fig. 1. City ontology – schematic approach

To simplify the notation, each set of parameterized shapes in the ontology is annotated as S_i where the index i defines the position of the set in the ontology from 1, 2, ..., n where n is the total number of shape sets in the ontology.

As defined above, an Urban Induction Pattern is a recurrent urban design move provided with a generic grammar that replicates this design move and can be applied independently of context. An Urban Grammar is a specific arrangement of generic grammars encoding urban design moves, i.e., it is an arrangement of UIPs.

The City Induction generation module was conceived of as an urban design tool or urban design system. In order to generate designs it contains a very generic urban grammar Γ which is the set of all urban grammars Γ' that can be defined using the system. Although our model is a simplified version of a possible theoretical universal model restricted to the domain defined by four case studies (see section 5), it can, in principle, be extended by (1) extending the ontology, and (2) extending the set of available UIPs.

An Urban Grammar Γ' is a Cartesian product of parallel grammars $\Gamma_1 \times \Gamma_2 \times \Gamma_3 \times \cdots \times \Gamma_n$ that use a set of parameterized shapes from the City ontology, $S_1, S_2, S_3, \cdots, S_n$ respectively, to design a layer of an urban plan. Each design phase outputs a sub-design composed of several layers or, more accurately, each design phase uses some of the grammars, Γ_1 to Γ_m, in the urban grammar Γ' to generate the various layers that define the sub-design produced in that design phase. Label sets $L_1, L_2, L_3, \cdots, L_n$, are the label sets in the corresponding grammars $\Gamma_1, \Gamma_2, \Gamma_3, \cdots, \Gamma_n$, and also correspond to the attribute classes in the ontology.

Generation begins with the definition of the existing elements to be used in the design. E_0 is the set of existing representations (shapes) of the working context and contains all the existing representations (shapes) regardless of the many-layered structure it may have. I_0 is the set of initial features (shapes and labels) that will be used to start the design and contains only I_s and R_{ef} objects, that is, the shape I_s, a closed polyline that represents the intervention site limit, the labels I_s and R_{ef}, and R_{ef} shapes, which are the shapes representing the selected elements in the site and context that will be used as referential guiding elements to support design rules. R_{ef} labels are attributes of the R_{ef} shapes that were selected by the designer to guide a certain stage of the design. R_{ef} shapes in fact represent the elements that guide the design decisions.

A grammar Γ' is built up from a sequence of UIPs (a sequence of design decisions) which in the end will reflect the design language of the urban plan. The same urban grammar Γ' can be used to generate different alternative designs by running the same UIP sequence again to produce different instantiations. This allows design implementations to be explored or monitored [Beirao, Duarte, Montenegro, Gil 2009]. The whole concept is, in fact, close to what could be a real algorithmic implementation of a complete Pattern Language as conceived of by Alexander [1977], but in such a way as to allow the designer to define his own pattern language.

A UIP is therefore a sub-grammar of Γ'. A UIP uses only some of the parallel grammars in Γ', a subset Γ'' of Γ', namely some components of $\{\Gamma_1, \Gamma_2, \Gamma_3, \cdots, \Gamma_n\}$. Each grammar Γ_i follows the definition of a discursive grammar [Duarte 2001]. Each UIP is a compound grammar Γ'' composed of a set of parallel discursive grammars Γ_i of the form $\Gamma_i = \{D, U, G, H, S_i, L_i, W, R, F, I_i\}$ where S_i is the set of parameterized shapes corresponding to the ith shape object class in the ontology, L_i is the set of labels corresponding to the ith attribute object class in the ontology and I_i is the initial shape.

The initial shape I_i is always a shape in S_i generated by a previous UIP or a shape in I_0 in the case of initial UIPs. Each UIP addresses a goal G to be achieved through set of description rules D starting from an initial description U. A set of heuristics H decides which of the rules in the set of rules R to apply. W is a set of weights and F a set of functions used to constrain generation so that it respects regulations and quality standards.

5 Generating designs with Urban Induction Patterns

In this section we intend to demonstrate that the concept described above can be used to design urban plans and explore design spaces. Since it would be impossible to start with the aim of defining all possible urban design moves, we used the following approach. The work was framed by capturing UIPs only within the design space defined by four case studies, attempting to define them in such generic terms that each UIP could be used as broadly as possible regardless of context. We also tried to break down the UIP into the smallest design moves possible so that most of the large design moves would already be a composition of smaller UIPs. In fact, the analysis of the case studies demonstrated that although all the grids were quite different in terms of final results, most of them were actually obtained through different arrangements of minimum UIPs designed for this purpose.

As a general approach, this section presents the four case studies, explains the approach used to infer UIPs, provides a summary of the UIPs inferred from the analysis and shows how these UIPs can be used (1) to replicate the urban plans, and (2) to generate alternative solutions by applying different instantiations of rule parameters.

The four case studies

This work began with the assumption that urban designs can be broken down into very small generic design procedures or design moves which, when combined in different ways, can be used to define different design languages. In order to define such design moves in the form of UIPs as defined in the previous section, we used four urban plans as case studies.

The four urban plans are: 1) Extension plan for Cidade da Praia in Cabo Verde by Chuva Gomes; 2) Qta da Fonte da Prata (QFP), in Moita, Portugal by Chuva Gomes; 3) Ijburg/Haveneiland by Frits van Dongen, Felix Claus and Ton Schaap from a larger plan by Palmbout; and 4) Ypenburg, also by Palmbout (Palmboom and van den Bout) (fig. 2). The case studies were used to frame the work within the range of their design space. Furthermore, we assumed that to produce complete urban designs we would have to define at least four sets of rules (UIPs) relating to four different levels of design:

A. rules to define the compositional guidelines of the plan;
B. rules to define grids or the main street structure;
C. rules to define urban units including squares and other public spaces;
D. rules for designing details, such as the detailed design of street profiles and materiality [Beirão and Duarte 2009].

This paper focuses on the level (B) rules.

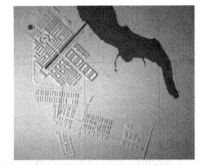
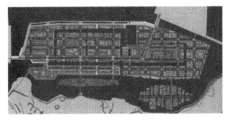

Fig. 2. Plans for (1) Praia, (2) Quinta da Fonte da Prata (both by Chuva Gomes), (3) Ijburg (by van Dongen, Claus and Schaap from a master plan by Palmbout), (4) Ypenburg (by Palmbout)

As a basic methodology we started by analysing the first case study, inferring the design rules used by the architect and defining them as Urban Induction Patterns. In interviews he has given, Chuva Gomes clearly states his design moves for both plans. He even states which moves were common and which were different in the two plans. We tried to formulate basic definitions of the UIPs in such a way that they would be valid for the four case studies or even other well known paradigmatic urban plans. Due to its simplicity, Case Study 1 provides very basic rules for designing an orthogonal grid-based plan. We tried to devise them so that, in order to obtain results like the ones in Case Study 2 the same rules (UIPs) could simply be applied with a different sequential arrangement and different values for the parameter variables. Roughly speaking, it can easily be seen that Plan 2 uses similar rules to those in Plan 1 but applies them several times in four different areas with four different orientations. Introducing more complexity into plans is therefore the result of combining basic UIPs in more complex ways. However, the two Dutch plans introduce new features suitable for encoding into other UIPs. For example, Plan 3 introduces a lot of variety into the definition of the urban block and a set of additional rules can therefore be defined from this case study. Case Study 4 introduces an even more complex variety of design transformations in comparison to the previous, less complex, case study. In this instance, not only can more variety be seen in terms of urban unit definition, but distortions in the orthogonal grid also become evident, implying the definition of rules to deal with grid distortions or irregular grids.

As a general approach, considering that our purpose is the definition of a design tool, the line of reasoning always followed was that rules to explain the simpler case study would be defined first and valid variations of these rules would then be explored as a means of obtaining the design exploration potential. This task proved hard, since it was possible to lose direction when considering valid variations simply due to the fact that design possibilities are, in principle, infinite. To avoid this, we tried to remain focused within the scope of the case studies, aiming only to prove that: (1) a minimum set of UIPs can be used to generate the four case studies; (2) the same UIPs can produce new designs in different contexts with similar rules in any language that can be composed from the UIPs available or, in other words, within the design space defined by the case studies. The additional capacities of the defined set of UIPs outside this design space were considered only as additional qualities to support the concept, but were not established as a research goal.

Defining UIPs

All the Urban Induction Patterns defined in this paper refer only to the generation of designs of urban plans in typical plan representations. 3D representations were left for future research developments and are considered here as consequence of the design generation results that might be obtained by simple extrusion of the layouts. Aspects related to topography are not detailed here, only clarified in the discussion.

As previously stated, the first UIPs applied are the ones that take the intervention site limit I_s and elements selected as references, R_{ef} as their initial shapes. The first UIP within the framework of our case studies was suggested by the explanation given by Frits Palmboom, the author of the Ypenburg plan, concerning his design methods. Regarding his first move in the design of an urban plan, in a 2008 interview made in the context of this present research, he said, "I always look for the longer line in the territory". Taking this sentence as our motto, we called the first UIP *MainAxisistheLongerLine*. The algorithm generates all the lines that can be generated after considering the referential elements and selects the longest one. It is applied in all the case studies and takes the particular form of the *Cardus* in the case of Plan 1, selecting from the proposed long lines the one that has the closest north-south orientation. The term *cardus* already demonstrates the possibility of applications in a wider design space than the one defined by the case studies, as it is a well-known feature of classic urban planning. Following the first UIP, we devised *OrthogonalAxis*, or *Decumanus* if perpendicular to *Cardus*, as another typical design move. In Plan 2 it can be seen that these two sequential UIPs are used differently in four different areas of the site. All of these are particular cases of *CompositionalAxis*, a UIP that uses two references to draw a compositional axis, and they are all related to the first design level (A). A detailed description of the rules for *Cardus* and *Decumanus* is shown in Beirão et al [2009a].

At the second design level (B) we basically considered two Urban Induction Patterns to generate the grids and a few others corresponding to complementary tasks that adapt the grid to predefined conditions, adjusting the results along the intervention boundaries or any other already existing element. The two UIPs generate grids by *AddingAxes* [Beirão, Duarte, Stouffs 2009b] or by *AddingBlockCells* [Beirão, Duarte, Montenegro, Gil 2009] and have been used to reproduce the design of Plan 1. The rules for each UIP are reproduced here to facilitate comprehension of this paper. However, some rules were added to the later UIP in order to deal with the use of varying parameters. During the derivation with *AddingBlockCells*, third hierarchy axes were adjusted using *AdjustingAxistoCells*. At the end of the grid derivation, two more UIPs were used – *AddBlocktoCells* and *AdjustingBlockCells* – to create the block structures and adjust to the boundary conditions of the design respectively.

As the architect Chuva Gomes used constant parameters to define the dimensions of the blocks in the Praia plan, the results of applying any of the two UIPs are the same. However, if the parameters were to vary, for instance, within a fixed interval, the results obtained from applying each of the algorithms would differ. Considering the goal of defining City Induction as a design tool, we view this difference as an extremely positive quality of the two UIPs, which can be used for design exploration.

Applying UIPs to generate designs

Exploring design possibilities is a task based on freely sketching many different ways of applying basic design moves to selected references within a given context. Both references and design moves may vary, as may the parameters or variables in each design move. Although the designer pursues a specific goal there are many possible solutions, even many optimum solutions, and the design exploration simply needs to find a way towards a solution space. Bearing this in mind, our aim is not necessarily to demonstrate the tool's capacity to find optimum solutions, although this may be achieved later when the three City Induction modules are connected, but to demonstrate the versatility of the design tool in exploring different design formalizations. In this sense, it is more important to show that, apart from being able to reproduce Chuva Gomes' plan, the same UIPs that were previously developed also enable many different solutions to be designed. In particular, it has to be demonstrated that different results can be obtained by: 1) selecting different references R_{ef}; 2) applying different sequences of UIPs; 3) assigning different values to the parameters.

Although some automated ways of developing suggestions for defining references can be found, until now this first step has simply been considered a manual one. The designer selects elements of the available representations to define as references for the design. References (R_{ef}) can be focal points (e.g., a hill top), lines or polylines (e.g., an existing street, a ridge, a water line) and polygons (e.g., a building). The concept of focal point is defined as a geometrical position with a tolerance corresponding to the tolerance designers use when sketching ideas in pencil or using any freehand tool. This means that the rules are structured to accept a certain flexibility with regard to the geometric position of the focal point. A building selected as a referential element (e.g., a historical building) can be treated by the rules as a focal point corresponding to its geometrical centre or as a polygon where the longer line is used either as an alignment or as the basis for establishing a perpendicular axis from its middle point. Different selections as well as different interpretations of the options will obviously produce different results.

Three initial UIPs are referred to in this paper – *MainAxisistheLongerLine*, *CompositionalAxis* and *Cardus*. Initial UIPs are those able to recognize the available initial shapes, which are the R_{ef} elements and the intervention site limit I_s. The basic algorithm for these three UIPs is the same and only the last step changes. They take all the selected R_{ef} elements and draw all the possible axes based on these elements. These axes are trimmed outside I_s. The longer axis defines the length l_{ax}, the longer axial length. From the whole set of axes only those 10% shorter than the longer axial length will be used in the next generation steps. However, this percentage is a variable that can be manipulated by the designer by changing the amount of proposed axes. *MainAxisistheLongerLine* selects the longest available axis from the set of proposed axes. *Cardus* selects the one closest to a north-south direction. *CompositionalAxis* randomly selects one of the available axes.[2] References, R_{ef} labels, used by the selected axes are erased so that a second coinciding axis cannot be generated. *MainAxisistheLongerLine* and *Cardus* can be applied only once. *CompositionalAxis* can be applied as long as there are references to be used.

OrthogonalAxis and *Decumanus* correspond to a second stage in UIP applications. They are applicable only if there is a main axis a_1 available or a *cardus*. *Decumanus* applies only if a *cardus* has been generated and only once in the whole design. *OrthogonalAxis* can be applied several times until there are no more references.

The first level representations generated by these UIPs are axial representations of streets belonging to the **AN** (Network) object class in the ontology and they basically represent four types of axes a_1, a_2, a_3 and a_4 corresponding to four distinct hierarchies. Other classifications can be added to the streets, detailing the street characteristics throughout the generation by adding attributes that change their configuration. In Plan 1 the architect decides to define the *decumanus* as a promenade and this feature is applied to all the a_2 axes in the plan, i.e., three times.

The UIPs *AddingBlockCells* and *AddingAxes* can be applied as soon as there are two orthogonal axes in the design. The parameters h and w correspond to the length and width of the urban block respectively. These parameters can be set as a fixed value, as Chuva Gomes does in Praia ($h=80m$ and $w=50m$), but can also be set as an admissible range (for instance: $60m \leq h \leq 120m$ and $40m \leq w \leq 100m$). In this case the results of applying these UIPs are different and their purpose becomes different in terms of design intentions.

Exploring variations in designs

At the beginning of the generation process the designer is prompted to define a minimum set of values that are used by the generation module as input values for specific parameters in the Urban Induction Patterns. In terms of the generation module these parameters can be set directly by the designer, although the formulation module is supposed to fill a table of specifications with such parameters as input data for the generation. In the next generation steps, in particular the exploration of grids, a few parameters must be defined, to be used by the rules in *AddingAxes* and *AddingBlockCells*. These parameters are the block length h and width w, and the street width defined for the hierarchy of compositional axes, a_1, a_2, a_3 and a_4. The latter widths can be altered during generation if an axis is transformed into a specific street type, for instance a *Promenade*, such as the three promenades found in Plan 1. However, the street widths are set as fixed values, while h and w can be set as a range of values. It is this permitted variability that makes the grid generator UIPs so interesting to explore.

We will look first at *AddingAxes*, as this is an easier example. [Beirão, Duarte, Stouffs 2009b] show the rules and the derivation for generating the Praia plan. The rules are reproduced here (fig. 4), showing a different derivation (fig. 3). The original derivation applies an exact sequence of UIPs and a fixed value for the block parameters, ($h = 80m$ and $w = 50m$), to reproduce the design of Praia plan – Plan 1. However, if we consider an admissible range for the parameters h and w, such as the ones suggested above, variations will start to appear in the grid whilst maintaining the typical orthogonal grid appearance and street continuity. The derivation in fig. 3 shows one possible solution resulting from the use of different values assigned to h and w randomly chosen from the stated range of values. No other function constrains the rule application in this example. At the end of the generation sequence, squares are applied following different algorithms for the generation of public space. The last Urban Induction Patterns apply two different building typologies to the block. These rules are not explained but their application is shown because it improves the legibility of the resulting urban plan.

Fig. 3 (this page and facing page). Derivation of the plan for Praia using AddingAxes. The derivation is simplified to the essential steps

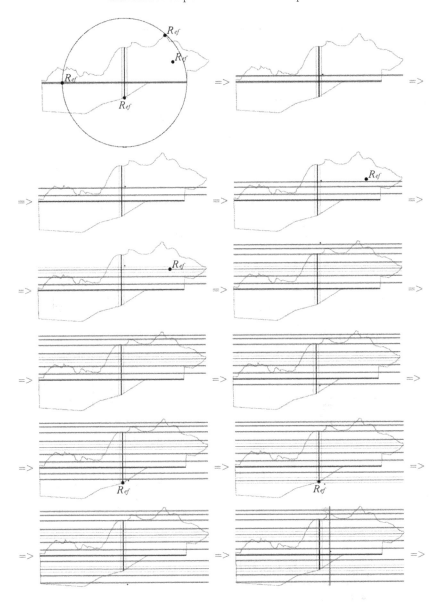

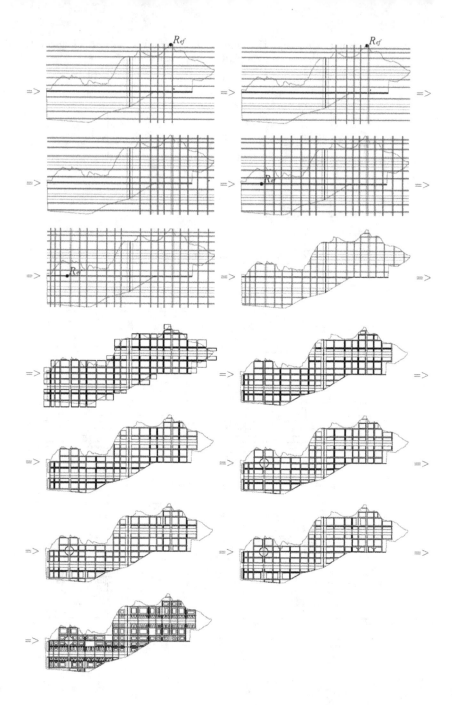

Fig. 4 (this page and facing page). Rules for AddingAxes (Rules 1, 2 and 7 are omitted)

UIP – *AddingAxes* – rules
Note: the dotted shapes in the rule representations are used only for reference.
No transformations are applied to the dotted shapes.
Parameters h and w are block length and block width respectively.
x and y are coordinates.
AddingAxes: $\Gamma''= \{\Gamma_1, \Gamma_0\}$

$$\{S_1, L_1\} \subset \Gamma_1 \,, \; S_1, L_1 \in \mathbf{AN}$$

$$\{S_0, L_0\} \subset \Gamma_0 \,, \; S_0, L_0 \in E_0$$

Rule 3a

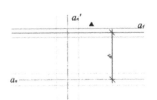

$\exists a_n, a_n{}' \in \mathbf{AN} \wedge a_n \perp a_n{}'$	$a_4 \in \mathbf{AN}$	$S_1 : \varnothing \to a_4 . y = d_2$
	$d_1 = w + a_{4\_width}/2 + a_{n\_width}/2$	$L_1 : \varnothing \to \blacktriangle$
$n \in \{1,2,3\}$	$\|a_4\| = |\min x, \max x|$	

Rule 4a

While $y < max_y$	$a_4 \in \mathbf{AN}$	$L_1 : \blacktriangle \to \varnothing$
	$d_2 = w + a_{4\_width}$	$S_1 : \varnothing \to a_4 . y = d_2$
	$\|a_4\| = |\min x, \max x|$	$L_1 : \varnothing \to \blacktriangle$

Rule 4c

If:	$a_4, a_n \in \mathbf{AN}$,	$L_1 : \blacktriangle \to \varnothing$
$\dfrac{1}{2}w - a_{4\_width} < p < \dfrac{3}{2}w + a_{4\_width}$	$n \in \{1,2,3\}$	$S_1 : \varnothing \to a_n . y = d_2$
	$d_2 = a_{4\_width}/2 + w + a_{n\_width}/2$	$I_{r_0} : R_{ef} \to \varnothing$
While $y < max_y$	$\|a_n\| = \|a_4\| = |\min x, \max x|$	$L_1 : \varnothing \to \blacktriangle$

Rule 5a

$\exists a_n, a_n' \in \mathbf{AN} \wedge a_n \perp a_n'$ $a_4, a_n, a_n' \in \mathbf{AN}$, $n \in \{1,2,3\}$ $S_1 : \varnothing \to a_4, x = d_3$

$d_3 = h + a_{4\_width}/2 + a_{n\_width}/2$ $L_1 : \varnothing \to \blacktriangle$

$n \in \{1,2,3\}$ $\|a_4\| = |\min y, \max y|$

Rule 6a

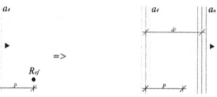

While $x < \max_x$ $a_4 \in \mathbf{AN}$, $n \in \{1,2,3\}$ $L_1 : \blacktriangle \to \varnothing$

$d_3 = h + a_{4\_width}$ $S_1 : \varnothing \to a_4, x = d_3$

$\|a_4\| = |\min y, \max y|$ $L_1 : \varnothing \to \blacktriangle$

Rule 6c

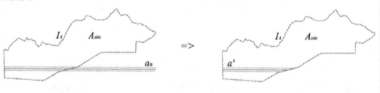

If

$\frac{1}{2} h - a_{4\_width} < p < \frac{3}{2} h + a_{4\_width}$ $a_4, a_n \in \mathbf{AN}$, $n \in \{1,2,3\}$ $L_1 : \blacktriangle \to \varnothing$

$d_4 = a_{4\_width}/2 + h + a_{n\_width}/2$ $S_1 : \varnothing \to a_n, x = d_3$

While $x < \max_x$ $\|a_4\| = |\min x, \max x|$ $L_0 : R_{ef} \to \varnothing$

$L_1 : \varnothing \to \blacktriangle$

Rule 8

$\forall a_n \in \mathbf{AN}$, $n \in \{1,2,3,4\}$ $a' = a_n \cap A_{site}$ $S_1 : a_n \to a'$

$a_n, a' \in \mathbf{AN}$

A_{site} is the region defined by the boundary I_s.

Apart from the fact that we let the rules randomly assign the h and w parameters, all the other steps in the design attempt to produce a fair replication of the Praia plan, so that the result underlines the difference in applying fixed or variable parameters to the block size. The derivation in fig. 3 was simplified to the essential steps.

AddingAxes is a sub-grammar Γ'' of the generic urban grammar Γ' that we are using. Γ'' is composed of two parallel grammars, Γ_1 and $\Gamma_0 - \{\Gamma_1, \Gamma_0\}$ – in which S_1, L_1 are the shape and label sets in Γ_1 respectively and both are objects of the **AN** class in the ontology. S_0, L_0 are the shape and label sets in Γ_0 respectively and they are both objects from set E_0.

The first step in the derivation already demonstrates the result of applying *Cardus*, *Decumanus* and a *Promenade* and shows all the R_{ef} points still to be erased. These R_{ef} points will be used to attribute a higher level of hierarchy to some of the axes generated. The second step starts the application of *AddingAxes*. Steps 4-5, 9-10, 15-16 and 18-19 show the steps where the R_{ef} points change the hierarchy of the axis to a higher level. Steps 4-5 and 9-10 apply a new *Promenade* to the axes passing through the first two of these referential points. Due to space restrictions, some of the repetitive steps were condensed. The last steps apply the generic blocks (UIP - *AddBlocktoCells*, step 20), adjust the blocks to the site boundaries (UIP - *AdjustingBlockCells*, step 21), create squares by subtracting some of the blocks (step 22), create the main plaza (step 23), create smaller squares by shrinking the block and reducing one of its parameters (step 24) or by subtracting some corners on a crossroad (step 25), and, finally, replace the generic blocks with two different types of building occupation, the closed block, composed of buildings surrounding the entire block, and a spine-like building occupation with the continuous side facing the main streets (step 26).

Fig. 4 shows the UIP *AddingAxes* with the rules 3a, 4a, 4c, 5a, 6a, 6c and 8. Some rules are omitted because they are symmetrical to others, namely rules 3b, 4b, 4d, 5b, 6b and 6d, which are symmetrical to rules 3a, 4a, 4c, 5a, 6a, and 6c respectively. *AddBlocktoCells* and *AdjustingBlockCells* are not shown in this paper but they can be found in an incomplete format in [Beirão, Duarte, Montenegro, Gil 2009].

Rule 1 of the UIP *AddingAxes* maps a temporary coordinate system, $x0y$, into two perpendicular axes a_n and a_n'. In this case, the axes are the ones generated by the patterns *Cardus* and *Decumanus*. Rule 2 extracts the maximum and minimum coordinate values from I_s taking the new coordinate system into account. The points are:

max_x = maximum x value of I_s in $x0y$;
max_y = maximum y value of I_s in $x0y$;
min_x = minimum x value of I_s in $x0y$; and
min_y = minimum y value of I_s in $x0y$.

These values will be used to frame the generation within the space defined by these coordinates.

Rule 3a adds a street axis – a_4 – parallel to a_1 or to the *cardus*. Labels ▲ and ▼ are used to define the recursive application of rule 4 (a, b, c and/or d) and indicate the direction in which to apply the next rule. Rule 3b is symmetrical to rule 3a and is applied in the negative y coordinate direction. Rule 4a adds a street axis – a_4 – parallel to an a_4 axis labelled with ▲, erases the label on the original a_4 axis and creates a new ▲ label on the new a_1 axis. The rule applies recursively until it falls outside the intervention site, i.e., while $y<max_y$ where y is the new a_1 y coordinate referred to by $x0y$. Rule 4b is symmetrical to rule 4a and is applied recursively in a similar fashion. If a selected R_{ef} point or element is within the region of a new axis, the designer is prompted to decide whether he wants to apply a new level of hierarchy to this new axis. This application is optional but if the designer does choose to use it, the axis is transformed into a higher level of hierarchy axis. Rule 4c exists for this purpose and rule 4d is symmetrical to it. Steps 2-11 in the derivation show the application of these rules.

Similar rules apply to the orthogonal axis a_2 or the *decumanus* to generate an array of perpendicular axes along the x coordinate. Rules 5a, 5b, 6a, 6b, 6c and 6d are used to generate these axes. Steps 12-19 in the derivation show the application of these rules.

The rules shown here are exactly the same as those used to generate the Praia plan, except that the values given to the parameters h and w are different in each iteration. The parameter values were defined randomly merely to explore design variations. However, this input could be informed through the formulation module, with specific values taking contextual data extracted from the site into account.

Rules 7a, 7b, 7c and 7d erase the ▲, ▼, ▶ and ◀ labels, respectively if they fall outside the framed area. Rule 8 trims the axes outside the I_s limit and rule 9 returns to the original coordinate system.

AddingAxes works with variable parameters in more or less the same way as it does with fixed values. On the other hand, *AddingBlockCells* behaves in an entirely different way. Defining each cell in the generation with different parameters creates a huge range of possible variations and a lot of unpredictability throughout the different steps of the derivation. In order to deal with this complexity, the set of rules in this UIP had to be expanded in comparison with the ones previously shown in [Beirão, Duarte, Montenegro, Gil 2009], in which the goal of the grammar was the replication of Praia plan.

Let us again assume that the generation will use an admissible interval for setting the values for the length h and width w of the block (again: $60m \leq h \leq 120m$ and $40m \leq w \leq 100m$). The block cell is defined by the block parameters plus the streets confining the block, which may be all different in some extreme cases (see fig. 5).

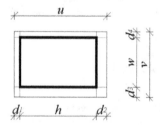

Fig. 5. The block cell – parameters and labels

Since each iteration can have different h and w values, the configurations of the design may contain many different variations, making recognition of the left hand side of the rules extremely difficult to manage.

Fig. 6 (this page and facing page). UIP AddingBlockCells – main rules

UIP – *AddingBlockCells* – **main rules**
Parameters h and w are block length and block width respectively. The coordinates x and y refer to the left bottom corner label + of the *void* area or to two main street crossings.
AddingBlockCells: $\Gamma" = \{\Gamma_1\}$

$$\{S_1, L_1\} \subset \Gamma_1, \; S_1, L_1 \in \mathbf{AN}$$

Rule 1

$\exists a_1, a_2 \in \mathbf{AN}$

a_1 and a_2 can also be
a cardus and a decumanus

$L_1 : \varnothing \to \bullet$ (4x)

Rule 2

$a_n, a_n', a_4 \in \mathbf{AN}$,

$n \in \{1, 2, 3\}$

$d_1 = a_{n'\_width} / 2$, $d_2 = a_{n\_width} / 2$,

$d_3 = d_4 = a_{4width} / 2$

$\|a_4\| = u = d_1 + h + d_4$ (hor)

$\|a_4\| = v = d_2 + w + d_3$ (ver)

$L_1 : \bullet \to \varnothing$

$S_1 : \varnothing \to a_4, x = u$ (ver)

$S_1 : \varnothing \to a_4, y = v$ (hor)

$L_1 : \varnothing \to \bullet$ (ver)

$L_1 : \varnothing \to \bullet$ (hor)

Rule 3

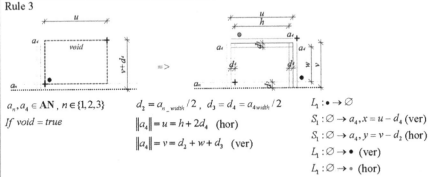

$a_n, a_4 \in \mathbf{AN}$, $n \in \{1, 2, 3\}$

If void = *true*

$d_2 = a_{n\_width} / 2$, $d_3 = d_4 = a_{4width} / 2$

$\|a_4\| = u = h + 2d_4$ (hor)

$\|a_4\| = v = d_2 + w + d_3$ (ver)

$L_1 : \bullet \to \varnothing$

$S_1 : \varnothing \to a_4, x = u - d_4$ (ver)

$S_1 : \varnothing \to a_4, y = v - d_2$ (hor)

$L_1 : \varnothing \to \bullet$ (ver)

$L_1 : \varnothing \to \bullet$ (hor)

Note: The + mark in the *void* vertices is used just to increase legibility and should not be read as part of the rules.

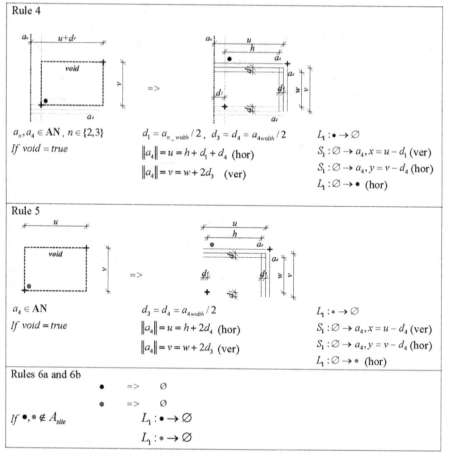

Rule 4

$a_n, a_4 \in AN$, $n \in \{2,3\}$

If *void* = *true*

$d_1 = a_{n\_width}/2$, $d_3 = d_4 = a_{4width}/2$

$\|a_4\| = u = h + d_1 + d_4$ (hor)

$\|a_4\| = v = w + 2d_3$ (ver)

$L_1 : \bullet \to \varnothing$

$S_1 : \varnothing \to a_4, x = u - d_1$ (ver)

$S_1 : \varnothing \to a_4, y = v - d_4$ (hor)

$L_1 : \varnothing \to \bullet$ (hor)

Rule 5

$a_4 \in AN$

If *void* = *true*

$d_3 = d_4 = a_{4width}/2$

$\|a_4\| = u = h + 2d_4$ (hor)

$\|a_4\| = v = w + 2d_3$ (ver)

$L_1 : \bullet \to \varnothing$

$S_1 : \varnothing \to a_4, x = u - d_4$ (ver)

$S_1 : \varnothing \to a_4, y = v - d_3$ (hor)

$L_1 : \varnothing \to \bullet$ (hor)

Rules 6a and 6b

$\bullet \quad \Rightarrow \quad \varnothing$

$\bullet \quad \Rightarrow \quad \varnothing$

If $\bullet, \bullet \notin A_{site}$

$L_1 : \bullet \to \varnothing$

$L_1 : \bullet \to \varnothing$

Liew points out three main problems in the use of shape grammars: "(1) controlling rule selection and sequencing in a grammar; (2) filtering out information in a drawing for rule application; and (3) specifying contextual requirements of a schema" [Liew 2004: 14]. He provides seven descriptors to be used in the rule application process to solve some of these problems. Specifically, with regard to contextual requirements he proposes the use of a descriptor "zone" which associates an area in a schema with a predicate function. He gives the example of a *void* function which states that a certain area must be empty of all shapes or specific shapes for the rule to apply. Because of the unpredictability of the *AddingBlockCells* grammar a similar descriptor needs to be used in its rules. The rule checks the context locally every time it is applied. The main rules for *AddingBlockCells* are basically the same six main rules as in [Beirão, Duarte, Montenegro, Gil 2009] but the descriptor zone was added to three of them (see fig. 6).

Rule 1 (fig. 6) places 4 labels • in the intersection of an a_1 and an a_2 axis, or in the intersection of a *cardus* and a *decumanus*. Rule 2 starts the cell derivation, erasing one of the labels • and adding two orthogonal a_4 axes. The cell width v and cell length u are defined by the values randomly chosen for w and h respectively, added to half the width of the streets that flank them. Rule 3 is applied recursively until the *void* predicate is no longer satisfied. The values for w and h are randomly chosen from the admissible range in each iteration. To ensure the recursive behaviour, rule 3 erases the original label ● and places another label ● next to the new a_4 axis on the right-hand side of the cell so that it can be used by the same rule in the following iteration. The rule creates a second label ● in the left top corner of the cell above the new a_4 axis, to be used later by Rule 5. Labels ● are only recognized by Rule 5 and their adjustment variations (as shown in fig. 7). To summarise, Rule 3 creates cells along the a_1 axis or any a_n axis parallel to the x coordinate where $n \in \{1,2,3\}$, until a vertical axis a_n' is found in the area checked by the *void* zone. There are 4 different situations that can occur if the *void* predicate is false. These 4 situations are the adjustment rules 3A_1, 3A_2, 3A_3 and 3A_4 (fig. 7). Rules 3A_1 and 3A_2 adapt the size of the new cell or the previous cell to meet the a_n' axis and create a new ● label on the right-hand side of a_n' to allow a new generation sequence to start. Conversely, Rules 3A_3 and 3A_4 move the a_n' axis until it fits the length u of the cell. These rules can be applied only if a_n is the main axis in the design or, in other words, if a_n was generated by *Cardus* or *MainAxisistheLongerLine*. This guarantees that an a_n' axis will not be moved more than once. Once again, a new label ● is placed on the right-hand side of a_n'.

Type A rules are all the adjustment rules that detect the presence of axes (objects from the **AN** object class) inside the *void* zone. Other types of adjustment rules react, for instance, to existing constructions, elements of set E_0, which are either streets or buildings. However, these rules are not shown here as there are no existing buildings in the Praia site.

Like the 3A rules, Rules 4A_1 and 4A_2 adapt the width v of the cell to meet the detected axis a_n' parallel to the main axis or the *cardus*. A label ● is created on the top of a_n' to allow another generation sequence to start in another area of the plan (see derivation in fig. 8). If a_n is a *decumanus* or the first orthogonal axis applied in the derivation, Rules 4A_3 and 4A_4 can be applied alternatively to adjust the position of a_n' to the cell size v. Like Rules 3A_3 and 3A_4, Rules 4A_3 and 4A_4 can be applied only once per axis.

While applying Rule 5 several occurrences may be detected inside the *void* zone. They are:

Rule 5A_1 and Rule 5A_4 detect the presence of one a_4 axis.

Rule 5A_2 and Rule 5A_3 detect the presence of one a_n axis.

Rule 5A_5 detects the presence of two a_n axes.

Rule 5A_6 and 5A_9 detect the presence of one a_4 axis and one a_n axis.

Rule 5A_7 detects the presence of one a_4 axis and two a_n axes.

Rule 5A_8 detects the presence of two a_4 axes and one a_n axis.

Rule 5A_1 detects the Δx length of penetration of a_4 axis inside the *void* zone and, depending on the value of Δx, produces two separate results. If $\Delta x \le u/2$, Rule 5A_1a generates a new cell creating two axes reducing the cell length u to u' so that $u' = u - \Delta x$. This rule creates a new ● label in the top left-hand corner of the cell. If $\Delta x > u/2$, Rule 5A_1b simply erases the existing ● label and creates a new one above it at a v distance in order to allow continuity of cell generation.

Fig. 7 (this page and following 4 pages). UIP AddingBlockCells (continuation) – adjustment rules.
When the void predicate is not satisfied, the main rules adjust to the conditions of the context

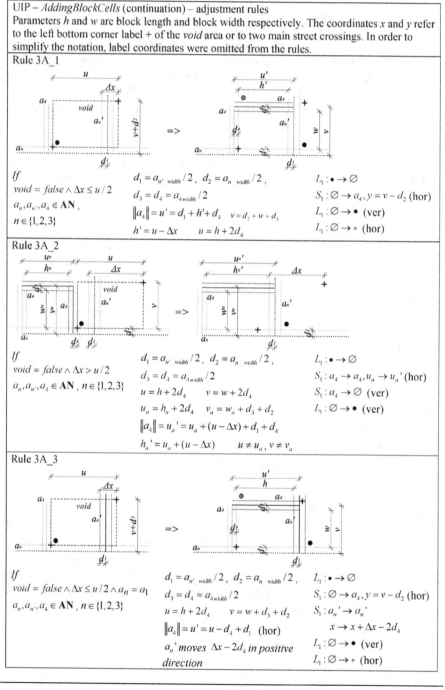

UIP – *AddingBlockCells* (continuation) – adjustment rules
Parameters h and w are block length and block width respectively. The coordinates x and y refer to the left bottom corner label + of the *void* area or to two main street crossings. In order to simplify the notation, label coordinates were omitted from the rules.

Rule 3A_1

If
void = false $\wedge$ $\Delta x \le u/2$
$a_n, a_{n'}, a_4 \in AN$,
$n \in \{1, 2, 3\}$

$d_1 = a_{n'\ width}/2$, $d_2 = a_{n\ width}/2$,
$d_3 = d_4 = a_{4\ width}/2$
$\|a_4\| = u' = d_1 + h' + d_4$ $v = d_2 + w + d_3$
$h' = u - \Delta x$ $u = h + 2d_4$

$l_1 : \bullet \to \varnothing$
$S_1 : \varnothing \to a_4, y = v - d_2$ (hor)
$l_1 : \varnothing \to \bullet$ (ver)
$l_1 : \varnothing \to \circ$ (hor)

Rule 3A_2

If
void = false $\wedge$ $\Delta x > u/2$
$a_n, a_{n'}, a_4 \in AN$, $n \in \{1, 2, 3\}$

$d_1 = a_{n'\ width}/2$, $d_2 = a_{n\ width}/2$,
$d_3 = d_4 = a_{4\ width}/2$
$u = h + 2d_4$ $v = w + 2d_4$
$u_a = h_a + 2d_4$ $v_a = w_a + d_3 + d_2$
$\|a_4\| = u_a' = u_a + (u - \Delta x) + d_1 + d_4$
$h_a' = u_a + (u - \Delta x)$ $u \ne u_a, v \ne v_a$

$l_1 : \bullet \to \varnothing$
$S_1 : a_4 \to a_4, u_a \to u_a'$ (hor)
$S_1 : a_4 \to \varnothing$ (ver)
$l_1 : \varnothing \to \bullet$ (ver)

Rule 3A_3

If
void = false $\wedge$ $\Delta x \le u/2 \wedge a_n = a_1$
$a_n, a_{n'}, a_4 \in AN$, $n \in \{1, 2, 3\}$

$d_1 = a_{n'\ width}/2$, $d_2 = a_{n\ width}/2$,
$d_3 = d_4 = a_{4\ width}/2$
$u = h + 2d_4$ $v = w + d_3 + d_2$
$\|a_4\| = u' = u - d_4 + d_1$ (hor)
$a_{n'}$ moves $\Delta x - 2d_4$ in positive
direction

$l_1 : \bullet \to \varnothing$
$S_1 : \varnothing \to a_4, y = v - d_2$ (hor)
$S_1 : a_{n'} \to a_{n'}$
$x \to x + \Delta x - 2d_4$
$l_1 : \varnothing \to \bullet$ (ver)
$l_1 : \varnothing \to \circ$ (hor)

Rule 3A_4

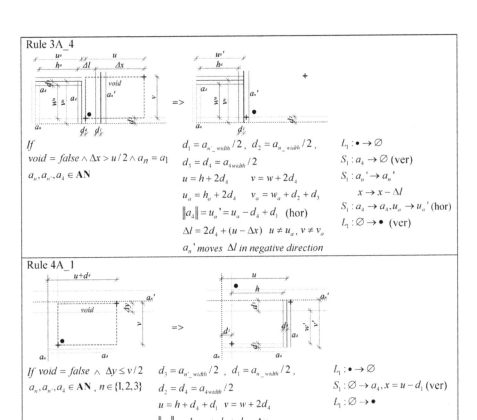

If
$void = false \wedge \Delta x > u/2 \wedge a_n = a_1$
$a_n, a_{n'}, a_4 \in \mathbf{AN}$

$d_1 = a_{n'\_width}/2$, $d_2 = a_{n\_width}/2$,
$d_3 = d_4 = a_{4width}/2$
$u = h + 2d_4 \qquad v = w + 2d_4$
$u_a = h_a + 2d_4 \qquad v_a = w_a + d_2 + d_3$
$\|a_4\| = u_a' = u_a - d_4 + d_1 \quad \text{(hor)}$
$\Delta l = 2d_4 + (u - \Delta x) \quad u \neq u_a, v \neq v_a$
$a_{n'} \text{ moves } \Delta l \text{ in negative direction}$

$L_1 : \bullet \to \varnothing$
$S_1 : a_4 \to \varnothing$ (ver)
$S_1 : a_{n'} \to a_{n'}$
$\qquad x \to x - \Delta l$
$S_1 : a_4 \to a_4, u_a \to u_a'$ (hor)
$L_1 : \varnothing \to \bullet$ (ver)

Rule 4A_1

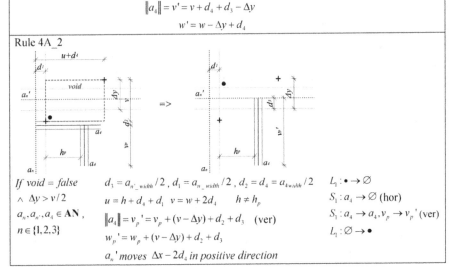

If $void = false \ \wedge \ \Delta y \leq v/2$
$a_n, a_{n'}, a_4 \in \mathbf{AN}$, $n \in \{1,2,3\}$

$d_3 = a_{n'\_width}/2$, $d_1 = a_{n\_width}/2$,
$d_2 = d_4 = a_{4width}/2$
$u = h + d_4 + d_1 \quad v = w + 2d_4$
$\|a_4\| = v' = v + d_4 + d_3 - \Delta y$
$w' = w - \Delta y + d_4$

$L_1 : \bullet \to \varnothing$
$S_1 : \varnothing \to a_4, x = u - d_1$ (ver)
$L_1 : \varnothing \to \bullet$

Rule 4A_2

If $void = false$
$\wedge \ \Delta y > v/2$
$a_n, a_{n'}, a_4 \in \mathbf{AN}$,
$n \in \{1,2,3\}$

$d_3 = a_{n'\_width}/2$, $d_1 = a_{n\_width}/2$, $d_2 = d_4 = a_{4width}/2$
$u = h + d_4 + d_1 \quad v = w + 2d_4 \quad h \neq h_p$
$\|a_4\| = v_p' = v_p + (v - \Delta y) + d_2 + d_3 \quad \text{(ver)}$
$w_p' = w_p + (v - \Delta y) + d_2 + d_3$
$a_{n'} \text{ moves } \Delta x - 2d_4 \text{ in positive direction}$

$L_1 : \bullet \to \varnothing$
$S_1 : a_4 \to \varnothing$ (hor)
$S_1 : a_4 \to a_4, v_p \to v_p'$ (ver)
$L_1 : \varnothing \to \bullet$

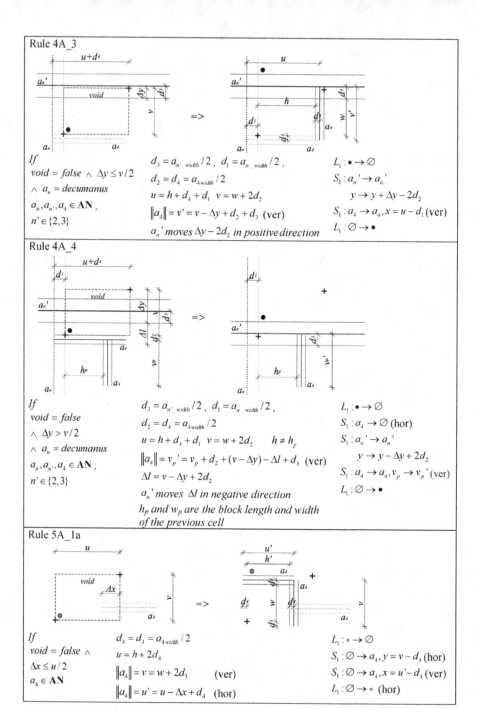

Rule 4A_3

$u+d_4$

a_n'

void

Δy d_3

v d_4

a_n a_4

$=>$

u

a_n'

h d_3

d_4 w v'

a_4

a_n a_4

If
$void = false \;\wedge\; \Delta y \le v/2$
$\wedge\; a_n = decumanus$
$a_n, a_{n'}, a_4 \in \mathbf{AN}$,
$n' \in \{2,3\}$

$d_3 = a_{n'\_width}/2\,,\;\; d_1 = a_{n\_width}/2\,,$
$d_2 = d_4 = a_{4width}/2$
$u = h + d_4 + d_1 \;\; v = w + 2d_2$
$\|a_4\| = v' = v - \Delta y + d_2 + d_3\;\;\text{(ver)}$
$a_n'\text{ moves } \Delta y - 2d_2 \text{ in positive direction}$

$L_1 : \bullet \to \varnothing$
$S_1 : a_n' \to a_n'$
$\quad\quad y \to y + \Delta y - 2d_2$
$S_1 : a_4 \to a_4, x = u - d_1\;\text{(ver)}$
$L_1 : \varnothing \to \bullet$

Rule 4A_4

$u+d_4$

d_1

void

Δy d_3

Δl d_2

v a_4

h_p

a_n a_4

$=>$

d_1

$+$

a_n'

d_3

d_2 v'

h_p a_4

a_n a_4

If
$void = false$
$\wedge\; \Delta y > v/2$
$\wedge\; a_n = decumanus$
$a_n, a_{n'}, a_4 \in \mathbf{AN}$,
$n' \in \{2,3\}$

$d_3 = a_{n'\_width}/2\,,\;\; d_1 = a_{n\_width}/2\,,$
$d_2 = d_4 = a_{4width}/2$
$u = h + d_4 + d_1 \;\; v = w + 2d_2 \quad\quad h \ne h_p$
$\|a_4\| = v_p' = v_p + d_2 + (v - \Delta y) - \Delta l + d_3\;\;\text{(ver)}$
$\Delta l = v - \Delta y + 2d_2$
$a_n'\text{ moves }\; \Delta l \text{ in negative direction}$
$h_p \text{ and } w_p \text{ are the block length and width}$
$\text{of the previous cell}$

$L_1 : \bullet \to \varnothing$
$S_1 : a_4 \to \varnothing\;\text{(hor)}$
$S_1 : a_n' \to a_n'$
$\quad\quad y \to y - \Delta y + 2d_2$
$S_1 : a_4 \to a_4, v_p \to v_p'\;\text{(ver)}$
$L_1 : \varnothing \to \bullet$

Rule 5A_1a

u

void

Δx v

a_4

$=>$

u'
h'

a_4

d_4 a_4

d_4 w d_4

v

a_4

If
$void = false \;\wedge$
$\Delta x \le u/2$
$a_4 \in \mathbf{AN}$

$d_3 = d_4 = a_{4width}/2$
$u = h + 2d_4$
$\|a_4\| = v = w + 2d_3 \quad\quad \text{(ver)}$
$\|a_4\| = u' = u - \Delta x + d_4\;\;\text{(hor)}$

$L_1 : \bullet \to \varnothing$
$S_1 : \varnothing \to a_4, y = v - d_3\;\text{(hor)}$
$S_1 : \varnothing \to a_4, x = u' - d_4\;\text{(ver)}$
$L_1 : \varnothing \to \bullet\;\text{(hor)}$

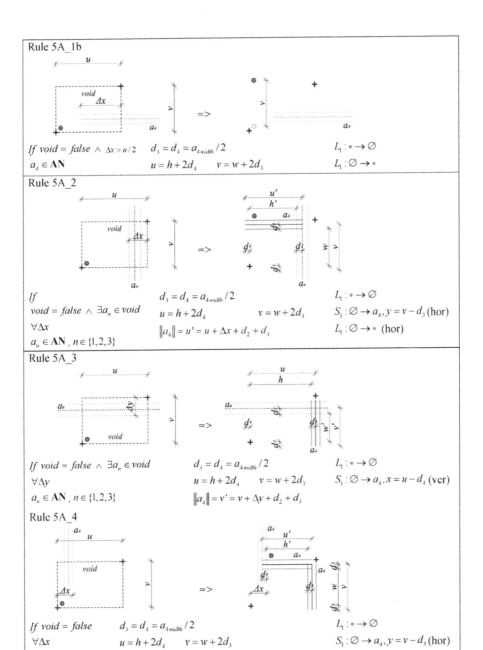

Rule 5A_1b

If $void = false \land \Delta x > u/2$ $d_3 = d_4 = a_{4width}/2$ $L_1 : \bullet \to \varnothing$

$a_4 \in \mathbf{AN}$ $u = h + 2d_4$ $v = w + 2d_3$ $L_1 : \varnothing \to \bullet$

Rule 5A_2

If

$void = false \land \exists a_n \in void$ $d_3 = d_4 = a_{4width}/2$ $L_1 : \bullet \to \varnothing$

$\forall \Delta x$ $u = h + 2d_4$ $v = w + 2d_3$ $S_1 : \varnothing \to a_4, y = v - d_3 \text{ (hor)}$

$a_n \in \mathbf{AN}, n \in \{1,2,3\}$ $\|a_4\| = u' = u + \Delta x + d_2 + d_4$ $L_1 : \varnothing \to \bullet \text{ (hor)}$

Rule 5A_3

If $void = false \land \exists a_n \in void$ $d_3 = d_4 = a_{4width}/2$ $L_1 : \bullet \to \varnothing$

$\forall \Delta y$ $u = h + 2d_4$ $v = w + 2d_3$ $S_1 : \varnothing \to a_4, x = u - d_4 \text{ (ver)}$

$a_n \in \mathbf{AN}, n \in \{1,2,3\}$ $\|a_4\| = v' = v + \Delta y + d_2 + d_3$

Rule 5A_4

If $void = false$ $d_3 = d_4 = a_{4width}/2$ $L_1 : \bullet \to \varnothing$

$\forall \Delta x$ $u = h + 2d_4$ $v = w + 2d_3$ $S_1 : \varnothing \to a_4, y = v - d_3 \text{ (hor)}$

$a_4 \in \mathbf{AN}$ $\|a_4\| = u' = u - \Delta x$ $h' = h - \Delta x$ $S_1 : \varnothing \to a_4, x = u - d_4 \text{ (ver)}$

   $L_1 : \varnothing \to \bullet \text{ (hor)}$

Rule 5A_5

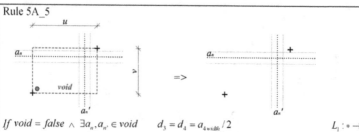

If void = false $\land$ $\exists a_n, a_{n'} \in void$ $\qquad d_3 = d_4 = a_{4\,width}/2$ $\qquad\qquad L_1 : \bullet \rightarrow \varnothing$

$$u = h + 2d_4 \qquad v = w + 2d_3$$

Rule 5A_6

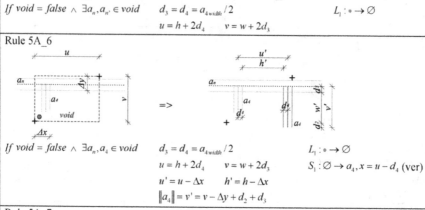

If void = false $\land$ $\exists a_n, a_4 \in void$ $\qquad d_3 = d_4 = a_{4\,width}/2$ $\qquad\qquad L_1 : \bullet \rightarrow \varnothing$

$$u = h + 2d_4 \qquad v = w + 2d_3 \qquad S_1 : \varnothing \rightarrow a_4, x = u - d_4 \text{ (ver)}$$
$$u' = u - \Delta x \qquad h' = h - \Delta x$$
$$\|a_4\| = v' = v - \Delta y + d_2 + d_3$$

Rule 5A_7

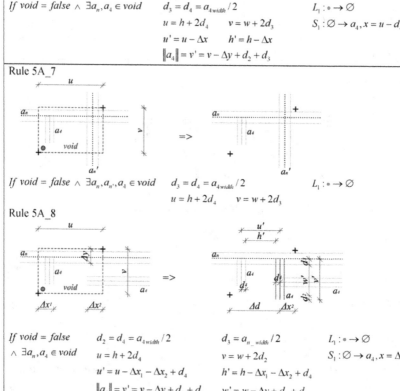

If void = false $\land$ $\exists a_n, a_{n'}, a_4 \in void$ $\qquad d_3 = d_4 = a_{4\,width}/2$ $\qquad\qquad L_1 : \bullet \rightarrow \varnothing$

$$u = h + 2d_4 \qquad v = w + 2d_3$$

Rule 5A_8

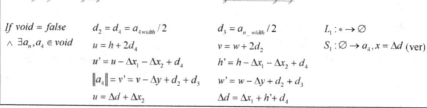

If void = false $\qquad d_2 = d_4 = a_{4\,width}/2$ $\qquad d_3 = a_{n\_width}/2$ $\qquad L_1 : \bullet \rightarrow \varnothing$

$\land$ $\exists a_n, a_4 \in void$ $\qquad u = h + 2d_4$ $\qquad\qquad v = w + 2d_2$ $\qquad\qquad S_1 : \varnothing \rightarrow a_4, x = \Delta d \text{ (ver)}$

$$u' = u - \Delta x_1 - \Delta x_2 + d_4 \qquad h' = h - \Delta x_1 - \Delta x_2 + d_4$$
$$\|a_4\| = v' = v - \Delta y + d_2 + d_3 \qquad w' = w - \Delta y + d_2 + d_3$$
$$u = \Delta d + \Delta x_2 \qquad\qquad \Delta d = \Delta x_1 + h' + d_4$$

Fig. 8 (this page and facing page). AddingBlockCells – derivation including adjustment rules

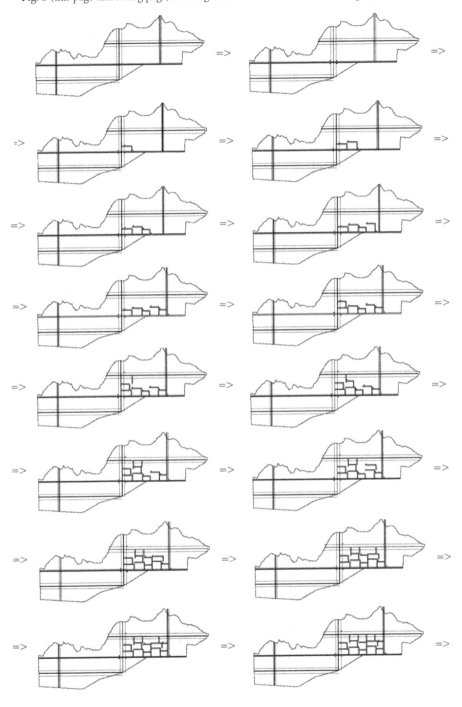

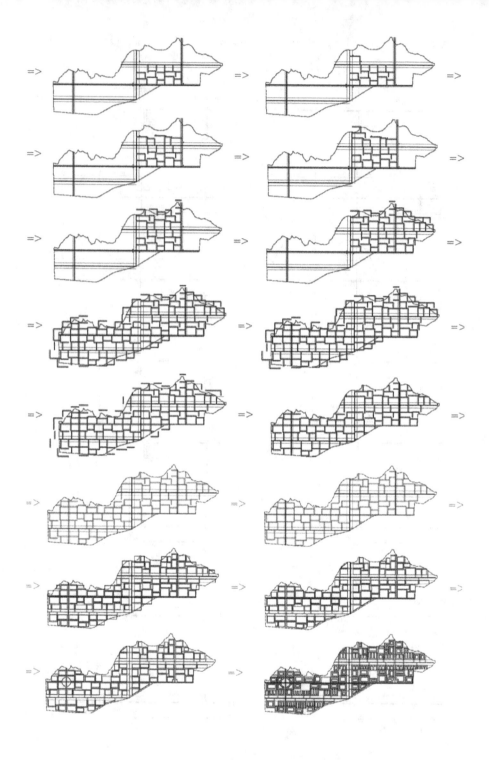

Rule 5A_2 reduces the cell size u to u' so that $u' = u - \Delta x + d_2 + d_4$ creates an a_4 axis closing the cell and a new label • in the top left-hand corner. The cell dimensions become $u' \times v$. Rule 5A_3 produces a similar result in the y coordinate direction, creating a cell with dimensions $u \times v'$ in which $v' = v - \Delta y + d_2 + d_3$. Note that this rule simply erases label • and does not create a new one.

Rule 5A_4 is similar to 5A_1 but it generates the reduced cell leaning against the right side of the void *zone* instead of the left side. In fact, it is using the free space in the *void* zone. u is reduced to u' through the relationship $u' = u - \Delta x$.

Rules 5A_5 and 5A_7 simply erase the label •. They are termination rules. Rules 5A_6 and 5A_8 generate a new cell, shortening both parameters u and v to u' and v' respectively. They both create an a_4 axis and do not recreate new • labels.

Rule 5A_9 reduces the cell length u to u' following the equation $u' = u - \Delta x_1 - \Delta x_2 + d_2 + d_4$, producing a cell with dimensions $u' \times v$.

The derivation in fig. 8 shows the generation of an urban plan using *AddingBlockCells* as the main algorithm for grid generation. The derivation starts in step 1, already demonstrating the result of applying *Cardus + Decumanus + 4 × OrthogonalAxis + 3 × Promenade*. Step 2 applies Rule 1 by placing 4 • labels. Step 3 applies Rule 2, generating the first cell, erasing one of the • labels and creating two more, associated with each of the newly created a_4 axes. Steps 4-6 apply Rule 3. Rule 3A_3 is applied in step 7. Note that a new • label is created on the right-hand side of the a_n' axis. Step 8 applies Rule 4 and step 9 applies Rule 4A_1, adapting the cell size v to the available circumstances. Steps 10 to 17 apply Rules 5 and 5A until all the • labels are erased. Generation is then terminated in this section of the plan. The other sections are generated in a similar fashion using the available • labels, starting with Rule 2 and ending with the exhaustion of labels • and •. Note that every label falling outside I_s is erased. Step 23 is the last cell creation step.

The whole set of rules allows the urban grid to be generated without conflicts but the final results still need adjustments, namely aligning or connecting a few streets and placing blocks within cells. However, these adjustments are produced by other UIPs. The derivation in fig. 8 clearly shows some of these problems in steps 23 to 26.

In step 24 the UIP *AlignStreets* is applied. The result of the generation using *AddingBlockCells* can produce several situations where a_4 axes connect to a_n axes that are very close to each other but not actually aligned. AlignStreets takes two axes connecting a higher level of hierarchy axis at points closer than three halves of their width and moves one of the axes to align with the other, creating a crossroads instead of a T junction. The choice of whether to move one or other of the a_4 axes depends only on the degree of connectivity of the axes. The axis with the least connectivity is chosen as the one to be moved. The criteria and degrees of connectivity used in this UIP are as follows, increasing from (1) to (3): (1) a_4 connects with another a_4 street, (2) a_4 has a corner connection with another a_4 and (3) a_4 aligns with another a_4 segment (see fig. 9).

Fig. 9 (this page and following 2 pages). UIP AlignStreets – Aligns two a4 streets when connecting to a higher hierarchy street. Moves the street with less connections or alignments in the street network

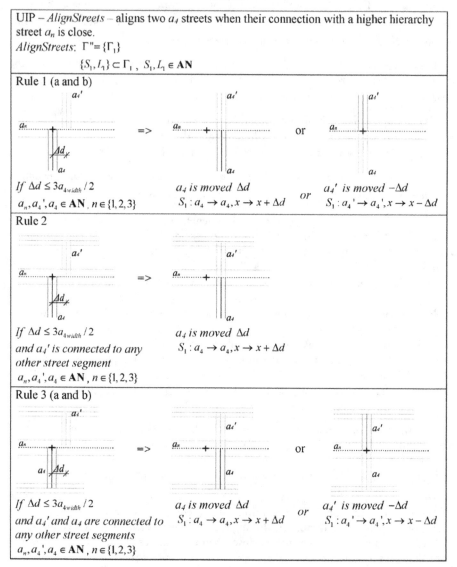

UIP – *AlignStreets* – aligns two a_4 streets when their connection with a higher hierarchy street a_n is close.

AlignStreets: $\Gamma'' = \{\Gamma_1\}$

$$\{S_1, L_1\} \subset \Gamma_1 \, , \; S_1, L_1 \in \mathbf{AN}$$

Rule 1 (a and b)

If $\Delta d \leq 3a_{4_{width}} / 2$

$a_n, a_4', a_4 \in \mathbf{AN} \, , \; n \in \{1, 2, 3\}$

a_4 is moved Δd

$S_1 : a_4 \rightarrow a_4, x \rightarrow x + \Delta d$

or

a_4' is moved $-\Delta d$

$S_1 : a_4' \rightarrow a_4', x \rightarrow x - \Delta d$

Rule 2

If $\Delta d \leq 3a_{4_{width}} / 2$
and a_4' *is connected to any other street segment*

$a_n, a_4', a_4 \in \mathbf{AN} \, , \; n \in \{1, 2, 3\}$

a_4 is moved Δd

$S_1 : a_4 \rightarrow a_4, x \rightarrow x + \Delta d$

Rule 3 (a and b)

If $\Delta d \leq 3a_{4_{width}} / 2$
and a_4' *and* a_4 *are connected to any other street segments*

$a_n, a_4', a_4 \in \mathbf{AN} \, , \; n \in \{1, 2, 3\}$

a_4 is moved Δd

$S_1 : a_4 \rightarrow a_4, x \rightarrow x + \Delta d$

or

a_4' is moved $-\Delta d$

$S_1 : a_4' \rightarrow a_4', x \rightarrow x - \Delta d$

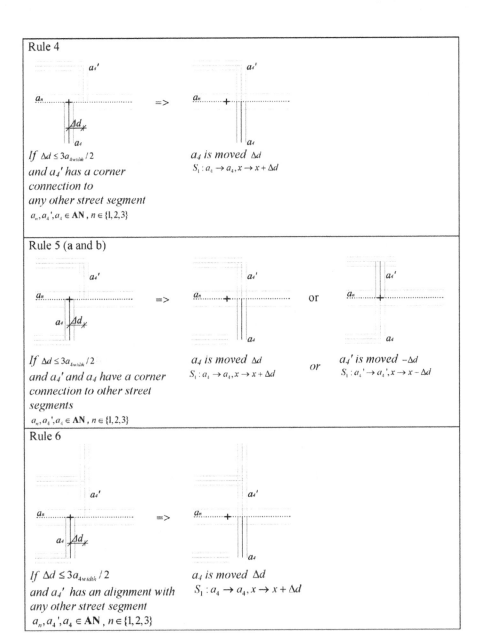

Rule 4

If $\Delta d \le 3a_{4\,width}/2$
and a_4' has a corner
connection to
any other street segment
$a_n, a_4', a_4 \in \mathbf{AN}$, $n \in \{1,2,3\}$

a_4 is moved Δd
$S_1 : a_4 \to a_4, x \to x + \Delta d$

Rule 5 (a and b)

If $\Delta d \le 3a_{4\,width}/2$
and a_4' and a_4 have a corner
connection to other street
segments
$a_n, a_4', a_4 \in \mathbf{AN}$, $n \in \{1,2,3\}$

a_4 is moved Δd
$S_1 : a_4 \to a_4, x \to x + \Delta d$

or

a_4' is moved $-\Delta d$
$S_1 : a_4' \to a_4', x \to x - \Delta d$

Rule 6

If $\Delta d \le 3a_{4\,width}/2$
and a_4' has an alignment with
any other street segment
$a_n, a_4', a_4 \in \mathbf{AN}$, $n \in \{1,2,3\}$

a_4 is moved Δd
$S_1 : a_4 \to a_4, x \to x + \Delta d$

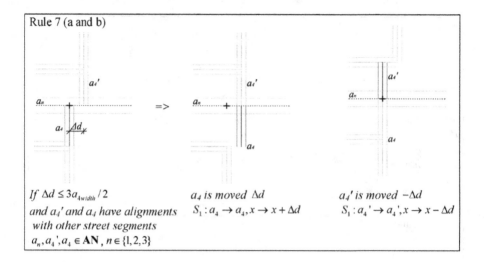

Rule 7 (a and b)

If $\Delta d \leq 3a_{4width}/2$
and a_4' and a_4 have alignments
with other street segments
$a_n, a_4', a_4 \in \mathbf{AN}$, $n \in \{1,2,3\}$

a_4 is moved Δd
$S_1 : a_4 \rightarrow a_4, x \rightarrow x + \Delta d$

a_4' is moved $-\Delta d$
$S_1 : a_4' \rightarrow a_4', x \rightarrow x - \Delta d$

In steps 25-26 all the axes falling outside I_s are either trimmed (step 25) or erased (step 26). Very small bits of a_4 axes are left from step 25. All the bits smaller than half of the lower value defined for h are erased (step 27). In this step it can be seen that some of the streets are still not connected and do not contribute towards the consistency of the street network. *ConnectStreets* is used to correct some of these inconsistencies in the grid generation (fig. 10). Step 28 shows the result of applying this UIP. In step 29 the cells are filled with abstract blocks using *AddBlocktoCells* [Beirão, Duarte, Montenegro, Gil 2009]. Step 30 adjusts or erases the blocks on the borders of the plan. Every block falling outside the I_s area is erased. In step 31 we can see the final result of applying several UIPs to square or public space generation. Steps 32-33 replace the abstract block with a few different block types showing the different possibilities for building occupation within the block. The last steps and UIPs are not detailed or discussed in this paper. However, the block types used in these rules are the same as the rules used by architect Chuva Gomes in Plan 1.

It is important to stress that it is not the intrinsic qualities or weaknesses of the final plan that are the goal in the given generation, but rather a demonstration of the versatility of the UIP *AddingBlockCells* in the generation of plans. This is accomplished simply by showing the results of randomly changing the h and w parameters for each iteration. The application of these parameters could be informed through other means such as the formulation module, in order to generate solutions following specific criteria.

Finally, we need to point out that the different UIPs only use objects from specific classes in the ontology. Likewise, the elements generated also belong to specific classes in the ontology and constitute different layered representations in the drawing. This structure is prepared for direct export to a GIS platform so that an integrated analysis can be performed within the context of the plan. In very general terms, all the axes belong to the **AN** object class, all the generic blocks belong to the **BL** object class and all ▲, ● and ● labels belong to the **AN** attribute class.

Fig. 10. UIP ConnectingStreets – Connects *a4* streets left unconnected after the *AddingBlockCells* and *AlignStreets* sequence

UIP – *ConnectStreets* Connects a_4 streets left unconnected after the *AddingBlockCells* and *AlignStreets* sequence.

ConnectStreets: $\Gamma'' = \{\Gamma_1\}$

$$\{S_1, I_1\} \subset \Gamma_1 , \ S_1, I_1 \in \mathbf{AN}$$

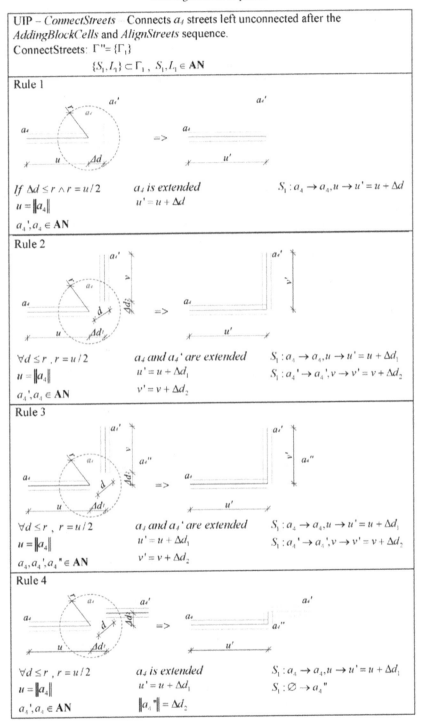

Rule 1

If $\Delta d \le r \wedge r = u/2$

$u = \|a_4\|$

$a_4', a_4 \in \mathbf{AN}$

a_4 is extended

$u' = u + \Delta d$

$S_1 : a_4 \to a_4, u \to u' = u + \Delta d$

Rule 2

$\forall d \le r , r = u/2$

$u = \|a_4\|$

$a_4', a_4 \in \mathbf{AN}$

a_4 and a_4' are extended

$u' = u + \Delta d_1$

$v' = v + \Delta d_2$

$S_1 : a_4 \to a_4, u \to u' = u + \Delta d_1$

$S_1 : a_4' \to a_4', v \to v' = v + \Delta d_2$

Rule 3

$\forall d \le r , r = u/2$

$u = \|a_4\|$

$a_4, a_4', a_4'' \in \mathbf{AN}$

a_4 and a_4' are extended

$u' = u + \Delta d_1$

$v' = v + \Delta d_2$

$S_1 : a_4 \to a_4, u \to u' = u + \Delta d_1$

$S_1 : a_4' \to a_4', v \to v' = v + \Delta d_2$

Rule 4

$\forall d \le r , r = u/2$

$u = \|a_4\|$

$a_4', a_4 \in \mathbf{AN}$

a_4 is extended

$u' = u + \Delta d_1$

$\|a_4''\| = \Delta d_2$

$S_1 : a_4 \to a_4, u \to u' = u + \Delta d_1$

$S_1 : \varnothing \to a_4''$

6 Discussion

The purpose of this paper is to show the versatility of design defined within the scope of the City Induction generation module by using an approach based on shape grammar. This is obtained at the level of grid generation simply by manipulating the range of admissible values for two parameters: the block length *h* and the block width *w*. The plan generation mechanisms are shape grammars encoding typical urban design moves. These grammars were called Urban Induction Patterns and designs are generated through the application of different arrangements of these UIPs. In this paper we show that the same rules used to generate one of the case studies – Plan 1 – can actually be used to generate other results, depending on how the available options in the system are manipulated, namely (1) to define a specific (sequential) arrangement of UIPs and (2) to manipulate the available rule parameters.

Current research is engaged in showing that a rather small set of UIPs can be used to generate a large amount of variety in design, including the scope of the four case studies. This paper already envisages that by using specific values for the parameters *h* and *w* for each iteration of *AddingAxes,* the grid of the Ijburg plan can be obtained (see plan 3 in fig. 2). In addition, a close look at Plan 2 shows that it uses approximately the same UIPs as Plan 1. This should not seem strange considering that the plans come from the same author but, nevertheless, careful examination shows that Plan 2 is divided into four sectors and each sector uses similar rule sequences. Only the parameters have considerably more variation. In each sector one main axis can clearly be seen with at least one orthogonal axis, a promenade and an orthogonal grid in which even some of the block types are similar. Hardly any new UIPs or new rules need to be added.

The interest of *AddingBlockCells* seems less evident in terms of grid generation, particularly since the example shown in fig. 8 looks a little unusual and maybe even a little messy. However, if we consider that instead of applying this UIP to highly structured top-down approaches such as the classic *cardus* and *decumanus* urban plans, the same or similar rules could be used to apply a bottom-up approach, i.e., generation starting from some focal point in an open area that progressively grows by adding cells step by step. This approach would definitely increase the interest of such an UIP, particularly if combined with different definitions for the urban units occupying the cells. Although some work is already being developed in this area, it certainly appears to be a promising research field in which there is still a lot to explore. Another important aspect is that it is evident that each grid generator UIP has a character of its own. *AddingAxes* generates a typical iron grid plan while *AddingBlockCells* generates a more informal urban fabric leaving some small public spaces with spatial characteristics that resemble certain areas in traditional cities. Furthermore, it can even be envisaged that both UIPs could be applied to different sectors of the plan, enabling a plan to be composed by developing areas with different characters.

The exploration of rules for block types is also clearly an interesting domain for further research. The last steps in the derivations shown here were not detailed but the rules are, in fact, very simple and can be roughly described as variations on the simplified rules shown in Figure 6 in Beirão et al [2008]. Pedro [1999] also suggests that most block occupations are variations or compositions based on only 3 types of block occupation: a closed block with a peripherical occupation, a linear block composed of parallel building blocks or an individual block composed of isolated buildings. The combination of these 3 types may show that most block compositions are actually based on very simple rules and the composition of very few basic elements. However, the Dutch case studies clearly show that block types can be very complex urban units and may give rise to the development of a few more UIPs to generate more complexity in the design at local level.

The main interest of this discussion is to support the hypothesis that a generation system based on the principles described in this paper can in fact be much simpler than initial examination may suggest. Future research will aim to prove this hypothesis, i.e., that a small set of UIPs with some available open parameters may be sufficient to provide a highly versatile design tool with generative capacities that can be easily integrated into GIS platforms. Furthermore, considering that design languages are personal expressions of designers, we would like to extend the goals of City Induction to provide ways of easily defining new UIPs or customizing existing ones in order to make not only an extremely generic, but also a highly customizable urban grammar available.

One of the most promising aspects of the use of the ontology which needs further exploration is the use of the relationships between the spatial components of the ontology through clear definitions of inheritance relations between object classes, relating these inheritance relationships to the UIPs in order to provide an automated means of reducing the set of available UIPs to be applied during the consecutive steps of the derivation without the need to establish a specific algorithm for this purpose. This is also related to the concept of a methodological approach to urban design which it is part of the task of this research to clarify. A future paper will address this subject.

3D representations have not been mentioned so far, but have not been forgotten. There are two different aspects to this subject. One simply concerns giving volume to the construction, which may easily be achieved by simply extruding the building polygons. This later needs to be linked to the generation process as the construction volume is directly related to urban parameters such as density, which is likely to be taken as a generation goal. The second aspect concerns topography and in this regard several cumulative issues might need to be taken into account. Two main categories of problem definitely emerge: (1) a methodological one, which may define a procedure to select the areas to be planned and urbanized by restricting the intervention area to a smaller zone or zones subject to distinct rules depending on slope, and (2) a rule-based approach to deal with topography and the design of streets and buildings in steep areas.

Finally, we need to remember that this paper focuses exclusively on the structure of the generation module. We should not forget that this is simply the design mechanism of a larger tool aimed at integrating programme formulation and evaluation techniques within the generation of designs. Additional functions emerging from the other two modules will improve the accuracy and credibility of the designs generated, as they may guide generation towards better solutions. It is planned to integrate the three modules supported by the said spatial ontology. However, this ontology still needs further detailing in order to integrate relationships between the morphological structures and the social and economical concepts used in formulation and evaluation processes. In addition, even at the morphological level further detailing is envisaged in order to approach the lower scale detailing of the plans.

7 Conclusions

Although it may be said that a design is the result of a sequence of design moves, it is not desirable to develop systems that impose a specific sequence for the simple reason that such decisions are context dependent. Urban Grammars, being an arrangement of generic grammars, allow for greater design flexibility in the sense that the sequence of procedures is not predefined but is instead defined by the designer during the design process. This paper demonstrates how a design tool is defined using generic grammars for the creation of customized urban designs produced by a customizable design language, a specific compound grammar defined by an arrangement of generic grammars.

The generation module for City Induction provides a very generic Urban Grammar composed of several generic grammars called Urban Induction Patterns corresponding to typical urban design moves. Specific grammars, such as the Praia grammar or the Ijburg grammar can be obtained by defining specific arrangements of Urban Induction Patterns and specific constraints on the rule parameters. As demonstrated in this paper, variations on the UIP arrangements or the rule parameters can provide design variations and allow for design exploration.

Acknowledgments

The City Induction project is funded by the Fundação para a Ciência e Tecnologia (FCT), Portugal (PTDC/AUR/64384/2006), hosted by ICIST at TU Lisbon, and coordinated by José P. Duarte. J.N. Beirão is also funded by the FCT by grant SFRH/BD/39034/2007. Beirão, J. Gil, and N. Montenegro are responsible for the generation, the evaluation, and the formulation modules, respectively. We would like to thank Henco Bekkering, Sevil Sariyildiz and Frank van der Hoeven for their readings, comments and support. Thank you to George Stiny and Terry Knight.

Notes

1. Zones indicating areas with specific features are represented in shape-files, as well as representations of real objects, but they are conceptual representations of our environment.
2. The use of weighted references has been considered as another possible way of selecting the axis instead of a random decision and it is closer to the real reasoning of a designer. The set of weights W is already considered for this purpose in the formal definition of a UIP. This principle is already being implemented in the software prototype.

References

ALEXANDER, Christopher, Sara ISHIKAWA and Murray SILVERSTEIN. 1977. *A Pattern Language: Towns, Buildings, Construction.* New York: Oxford University Press.

ASCHER, F. 2001. *Les nouveaux principes de l'urbanisme. La fin des villes n'est pas à l'ordre du jour.* La Tour d'Aigues: Éditions de l'Aube.

BATTY, M. 2005. *Cities and Complexity: Understanding Cities with Cellular Automata, AgentBased Models, and Fractals.* Cambridge, MA: MIT Press.

BEIRÃO, José Nuno and José Pinto DUARTE. 2009. Urban Design with Patterns and Shape Rules. Pp. 148-165 in *Model Town: Using Urban Simulation in New Town Planning,* EH Stolk and M Brömmelstroet, eds. 2nd International Seminar, 2007, Almere. Almere, The Netherlands: Martien de Vletter.

BEIRÃO, José; José Pinto DUARTE and Rudi STOUFFS. 2008. Structuring a Generative Model for Urban Design: Linking GIS to Shape Grammars. Pp. 929-938 in *Architecture 'in Computro'* (Proceedings of the 26th eCAADe Conference, Antwerp, 17-20 September 2008), Marc Muylle, ed. Antwerp.

———. 2009a. Grammars of designs and grammars for designing. In *Joining Languages, Cultures and Visions: CAADFutures 2009,* T. Tidafi and T. Dorta, eds. Montreal: University of Montreal.

http://cumincad.scix.net/cgi-bin/works/Show?cf2009_890

———. 2009b. An Urban Grammar for Praia: Towards Generic Shape Grammars for Urban Design. Pp. 575-584 in *Computation: The New Realm of Architectural Design* [Proceedings of the 27th eCAADe Conference, Istanbul, 16-19 September 2009. Istanbul: eCAADe, YTU, ITU.

BEIRÃO, José N., José Pinto DUARTE, Nuno MONTENEGRO and Jorge GIL. 2009. Monitoring urban design through generative design support tools: a generative grammar for Praia. Pp. 1223-1252 in *Proceedings of the 15th APDR Congress on Networks and Regional Development.* Cidade da Praia Cape Verde: APDR.

BEIRÃO, José Nuno; Nuno MONTENEGRO; Jorge GIL; José P. DUARTE; Rudi STOUFFS. 2009. The city as a street system: A street description for a city ontology. Pp. 132-134 in *SIGraDi 2009 - Proceedings of the 13th Congress of the Iberoamerican Society of Digital Graphics*, Sao Paulo, Brazil, November 16-18, 2009. São Paulo: eCAADe, Universidade Presbiteriana MacKenzie.

BROWN, F. E. and JOHNSON J. H. 1984. An interactive computer model of urban development: the rules governing the morphology of mediaeval London. *Environment and Planning B: Planning and Design* **12**, 4: 377-400.

DUARTE, José Pinto 2001. Customizing Mass Housing: a discursive grammar for Siza's Malagueira house. Ph.D. Dissertation, Massachusetts Institute of Technology, Cambridge.

DUARTE, José Pinto, J. ROCHA, G. DUCLA-SOARES. 2007. Unveiling the structure of the Marrakech Medina: A Shape Grammar and an Interpreter for Generating Urban Form. *AI EDAM Artificial Intelligence for Engineering Design, Analysis and Manufacturing* **21**: 1-33.

FLEISCHER, A. 1992. Grammatical architecture? *Environment and Planning B: Planning and Design* **19**, 2: 221-226.

FRIEDMAN, A. 1997. Design for Change: Flexible Planning Strategies for the 1990s and Beyond. *Journal of Urban Design* **2**, 3: 277-295.

GAMMA, Erich, Richard HELM, Ralph JOHNSON and John VLISSIDES. 1995, *Design Patterns: Elements of Reusable Object-Oriented Software.* Reading, MA: Addison-Wesley.

GIL, Jorge. and DUARTE, José Pinto. 2008. Towards an Urban Design Evaluation Framework. Pp. 257-264 in *Architecture 'in Computro'* (Proceedings of the 26th eCAADe Conference, Antwerp, 17-20 September 2008), Marc Muylle, ed. Antwerp.

HILLIER, Bill. 1998. Space is the Machine : A Configurational Theory of Architecture, Cambridge University Press.

HILLIER, Bill and Julienne HANSON. 1989. *The Social Logic of Space.* Cambridge: Cambridge University Press.

KNIGHT, Terry Weissman. 2003. Computing with ambiguity. *Environment and Planning B: Planning and Design* **30**, 2: 165-180

LIEW, Haldane. 2004. SGML: A Meta-Language for Shape Grammars. Ph.D. Dissertation, Massachusetts Institute of Technology.

MAYALL, Kevin and G. Brent HALL. 2005. Landscape grammar 1: spatial grammar theory and landscape planning. *Environment and Planning B: Planning and Design* **32**, 6: 895-920.

MONTENEGRO, Nuno C. and José Pinto DUARTE. 2008. Towards a Computational Description of Urban Patterns. Pp. 239-248 in *Architecture 'in Computro'* (Proceedings of the 26th eCAADe Conference, Antwerp, 17-20 September 2008), Marc Muylle, ed. Antwerp.

PARISH, Yoav I. H. and Pascal MÜLLER. 2001, Procedural modeling of cities. Pp. 301–308 in *Proceedings of ACM SIGGRAPH 2001*, E. Fiume, ed. New York: ACM Siggraph.

PEDRO, J. Branco. 1999. *Vizinhança Próxima* (Housing program. Neighbourhood). Lisbon: LNEC. (Collection Architecture Technical Information, n.º 7).

PORTUGALI, Juval. 1999. *Self Organization and the City.* Heidelberg: Springer.

SCHÖN, Donald A. 1983. *The Reflective Practitioner: How Professionals Think in Action.* New York: Basic Books.

STINY, George. 1980a. Introduction to shape and shape grammars. *Environment and Planning B: Planning and Design* **7**, 3: 343-351.

———. 1980b. Kindergarten grammars: designing with Froebel's building gifts. *Environment and Planning B: Planning and Design* **7**, 4: 409-462.

————. 1981. A note on the description of designs. *Environment and Planning B: Planning and Design* **8**, 3: pp. 257-267.

————. 2005. Shape, talking about seeing and doing, Cambridge, MA: MIT Press.

STINY, George and James GIPS. 1972. Shape Grammars and the Generative Specification of Painting and Sculpture. *Information Processing* **71**: 1460-1465.

TEELING, Catherine. 1996. Algorithmic Design: Generating Urban Form. *Urban Design Studies* 2: 89-100.

YUE, K., R. KRISHNAMURTI and F. GROBLER. 2009. Computation-friendly shape grammars: Detailed by a sub-framework over parametric 2D rectangular shapes. Pp. 757-770 in *Joining Languages, Cultures and Visions: CAADFutures 2009*, T. Tidafi and T. Dorta, eds. Montreal: Les Presses de l'Université de Montreal.

About the authors

José Nuno Beirão was awarded a professional degree in architecture from the Faculty of Architecture of the Technical University of Lisbon in 1989. He has practiced architecture and urban design since then. After working at the Falcão de Campos and Gonçalo Byrne Offices, he started the architectural firm B Quadrado Arquitectos with Miguel S. Braz in 1998. Their work and portfolio is available at www.bquadrado.com. José Nuno Beirão is now working towards his PhD dissertation at the TU Delft Faculty of Architecture. The research follows the subject of his Master's thesis ("Urban Grammars: Towards Flexible Urban Design," ISCTE - Instituto Superior das Ciências do Trabalho e da Empresa, Lisbon, 2005) and is part of a larger research project called City Induction hosted at the TU Lisbon. His research interests are the development of customizable and flexible design systems, and have focused on housing since 1998 and more intensively on urban design since 2001. His current interests are focused on the development of shape grammars for urban design and on the use of the generative capabilities of shape grammars to support the urban design process and foster design exploration.

José Pinto Duarte holds a B.Arch. (1987) in architecture from the Technical University of Lisbon and an S.M.Arch.S. (1993) and a Ph.D. (2001) in Design and Computation from MIT. He is currently Visiting Scientist at MIT, Associate Professor at the Technical University of Lisbon Faculty of Architecture, and a researcher at the Instituto Superior Técnico, where he founded the ISTAR Labs - IST Architecture Research Laboratories. He is the co-author of *Collaborative Design and Learning* (with J. Bento, M. Heitor and W. J. Mitchell, Praeger 2004), and *Personalizar a Habitação em Série: Uma Gramática Discursiva para as Casas da Malagueira* (Fundação Calouste Gulbenkian, 2007). He was awarded the Santander/TU Lisbon Prize for Outstanding Research in Architecture by the Technical University of Lisbon in 2008. His main research interests are mass customization with a special focus on housing, and the application of new technologies to architecture and urban design in general.

Rudi Stouffs is Associate Professor of Design Informatics and leader of the Computation & Performance research group and programme at the Faculty of Architecture, Delft University of Technology. He holds an MSc in architectural engineering from the Vrije Universiteit Brussel, an MSc in computational design and a Ph.D. in architecture from the Carnegie Mellon University (CMU). He has been an Assistant Professor in the Department of Architecture at the CMU and Research Coordinator of the Architecture and CAAD course at ETH Zurich. His research interests include computational issues in description, modelling, and representation for design in the areas of information exchange, collaboration, shape recognition and generation, geometric modelling, and visualization.

Mine Özkar

Department of Architecture,
Faculty of Architecture
Middle East Technical
University
Ankara, Turkey 06531
ozkar@metu.edu.tr

Keywords: design analysis, design
theory, shape grammars, shapes,
symmetry groups

Research

Visual Schemas: Pragmatics of Design Learning in Foundations Studios

Presented at Nexus 2010: Relationships Between Architecture
and Mathematics, Porto, 13-15 June 2010.

Abstract. Visual schemas are prequels to shape rules in
formalizing the pedagogical discussions in foundational
design studios.

The importance of talking in the foundations studio

It may well be true that if one half of the conduct of a design student in the studio is designing, the other half is talking about designing. The ideal studio environment allows for the students to vocalize to themselves as much as to others, the questions and the possible answers they have regarding their thinking and designing processes. It is important that the students in any design studio discuss what they are doing to improve and impede their learning. This is a way to vocalize the *reflection-in-action* and is even more significant in a foundations studio where the reflection habits are being established for the first time for the purposes of designing.

The view that design is to be talked about, explicitly, is increasingly shared among educators who acknowledge the growing role of process over the product in technologically integrated design processes. Since the beginnings in the early 1900s [Özkar 2005a], the issue of *how to* has been proposed as taking precedence over *what to* in design education. Despite heated debates, the interest and the conviction in explicit design processes accelerated with the Design Methods movement [Broadbent 1979] and continued with the notion of diagrams [Do and Gross 2001]. Increasingly integrated with information technologies, the contemporary paradigm of architectural design calls for open and dynamic processes of design. Today, more design ideas and processes are depicted through various, perhaps unclear but nevertheless seductive and stimulating diagrams and are encouraging for the growing field of inquiry into design methodologies and techniques.

Similarly in a growing number of design studios, methods that enable actions and critical discussions regarding the product are coming to the forefront. There is a good volume and breadth of literature regarding the various technical, social, pedagogical and practical issues surrounding the state of the art in design education, some with a focus on the studio. To mention a few key references, Findeli [2001] provides a comprehensive discussion of multiple aspects of the general topic while Oxman [2008] gives an accurate and extensive account of the impact of digital technologies on design education, and Stiny [2006] introduces a computable connection between *doing* and *talking* in design. Looking from the viewpoint of integrating computer technologies to design curricula as enablers of contemporary design processes, this paper assumes that the studio, particularly the foundations studio, is the medium not only for learning to produce but also for learning to talk about design. The purpose of talking in the foundations studio is twofold: one, understanding and developing design skills, all the while connecting them with the general notions of computation, in order to liberate thinking/reasoning from the association with quantitative and hierarchic structures alone; two, as part of a holistic view of design and design education where the student (the novice designer) is at the center as the actor of learning.

Beginning architecture students often display discouragement in discussing what is, to them, internal, unclear, and sometimes accidental in the work they produce. They often come with preconceptions of how designers behave based on the common market imagery of the architect and/or architecture, which is lacking in social and rational groundedness. Encouraging students to talk about what they do is a significant step for de-mystifying the notion of design in their eyes. Talking changes the image of the designer into an accessible one and shows that design is to be criticized based on various valid criteria, and is learnt and developed. The first year in architecture education is where the culture and habits of design thinking start to form and where the students are introduced to its tools for the first time. As tools of representation and thinking are constantly revised with the changing technology, it is necessary, for the sustainability of a contemporary design education, that students acquire from early on an awareness of the dynamic nature of design knowledge.

There is a growing literature on the student/learner-centered higher education paradigm, which is quite relevant to any scholar engaged in curricula or course design. Brown [2003] recapitulates the necessary conditions for it and the psychological factors involved. Although well beyond the scope of this paper, some of these factors and conditions, such as active and hands-on learning, student motivation and social development fall in line with what is epitomized here in talking. Talking in the studio involves asking questions, playing different roles, comparing, seeing, and doing something just to try it out as actions of self-reflection in a process of production. The studio environment is often considered as the ideal setting for a learner-centered higher education paradigm, but every studio setting does not by default fit this model. The key is in the attitudes and the discussion environment provided for the student to self-reflect and develop his or her own position in action. Where enabling the student as an active and self-critical subject is a key matter of concern, grounding studio learning on dialogue (through questions, panel discussions, critiques, etc.) is a common starting point. Nevertheless, in addition to just words, formalisms are needed for this conversation to be linked with visual and spatial thought processes in question.

Means for talking, formally, about basic design

In basic design education, a format for foundations studios in architecture schools throughout the world, tasks are often deemed abstract and removed from real design problems. First and foremost, the learning outcome of basic design is the design thinking experience and accordingly, the ability to relate various forms at different levels of complexity. Defining relations (some arrangement of what is conceived as separate parts) is usually a more complex task when there are fewer constraints in how to perceive the parts and wholes visually, as there are multiple levels and alternatives in perception.

Talking in the studio does not have to be formal but, formalisms are handy to follow, repeat and document the talking. The premise here is that basic design exercises are visual/spatial computations [Özkar 2005b]. Ideally, showing that their design works as a visual computation helps the student to observe what they are doing, to accommodate it within one's own sense and to develop it further. Basic design exercises seem especially suitable for observation in shape computation formalism, as they deal with less complex forms and relations. However, as basic design exercises display their own level of complexity due to their abstract nature, conveying the formalism of visual rules to the beginning student proves to be not as straightforward. This paper intends to draw attention to the practicality of the matter by showing what kinds of formalisms are relevant to the students' discussions in the context of basic design.

Shape grammars have been presented and widely used as devices of talking about design. Many scholars have worked towards integrating the formalism of shape grammars to design education. At various levels of architecture education, the formalism has been valuable for enabling the students to recognize and develop further the systemic, recursive aspects of their design, as well as the generative aspects of rule-based systems and parametric assemblies towards contemporary production processes. Knight [2000] has been one of the key promoters in architecture and has been diligently encouraging architecture students to bring their design processes to terms with visual rules. While Celani [2002] emphasizes recursion in design in her valuable attempts to introduce the shape grammar formalism to her students early on, Sass [2005] incorporates rule-based processes with fabrication. Additionally, there is a wide variety of tools that are developed to provide an interface with the shape grammars formalisms [Chase 2005].

The general idea for integrating the shape grammar formalism in the foundations studio is to incorporate shape rules as well as the knowledge of visual algebras and Boolean operations into the vocabulary of studio talk, in support of the conventional panel discussions. Hypothetically, similar to pointing at the visual composition with a finger and elaborating a visual relation with words, using rules should be an efficient and shared way to talk about design decisions. However, a few attempts quickly show that for any basic design there are too many visual relations to consider for shape rules. The abundance is confusing and futile unless one also talks of some higher-level relations between the shape rules. Then, the design knowledge to be conveyed seems to concern the relations of parts and wholes, not just at one level but at multiple levels. The next sections will be dedicated to expand on this design knowledge and how it can be turned into the matter of formal talk. All the while, the paper proposes that visual schemas come in handy to talk about this particular design knowledge.

What is meant by a schema here is a generalized version of a rule [Stiny 2006, 2009]. Visual schemas are more general ways to understand design decisions, processes and methods, whereas a visual rule can have more specifications. In this paper schemas are sometimes represented in abstract ways such as x→y but still refer to visual schemas as they deal with shapes. Drawing from these illustrations, the proposal is that visual schemas provide a more suitable level of discussion than the detailed shape rule formalism for foundations studio talk. As part of a preliminary study for incorporating formal methods in the foundations studio, this paper illustrates some visual schemas in two completed basic design examples based on what the students have explored and identified in their work without the knowledge of visual schemas.

Recurrence, emergence and visual schemas: foundational design knowledge?

In the keynote lecture at the 2009 eCAADe meeting, George Stiny emphasized once again the indispensible connection between recursion and embedding in the context of design computing. In addition to the usual coupling of recursion with the identity relation in symbolic computation, Stiny proposes that in design computation, part relations, specifically embedding, are indispensable partners to recursion. Recursion and embedding constitute a significant part of design knowledge in the context of computation and have been playing their parts, in the form of rule-based systems and emergence, in numerous practices of visual computation in design. Stiny's emphasis was not a mere recursion but also a reminder to see the simultaneity of the two aspects. This paper proposes that these two notions are part of the foundational design knowledge to be primarily conveyed in basic design exercises and that visual schemas are suitable for doing so.

Recursive functions in design computing have been acknowledged before [Stiny and Gips 1972; Stiny 1985; Kirsch and Kirsch 1986]. Existing examples of shape grammar applications in education since have mostly explored the recursive aspects of design via additive approaches, starting out with initial shapes set in some spatial relation which are iterated in space as relevant addition rules of the type x→ x+t(x) are applied. Basic design exercises, as abstract compositions, are not unlike these processes. They inevitably rely on recursion – or rhythm, to use a more designerly term – as it establishes links of similarity or variance between the parts, which in turn help the eye in reading what makes up the unity in a composition.

It is useful to identify the two common ways, one additive, another subtractive, of setting up basic design problems. An example of the additive approach would be something along the lines of "organize the given shapes in space," while an example of the subtractive approach would be something along the lines of "carve shapes out of a solid material." If the goal of a unified whole is also given, both approaches are usually treated as composition. However they differ greatly in the level of plasticity they provide. Compositions with primitives mostly tend to be orthographic (easily measured, geometrized and cut) and carving seems to promote free hand movements.

Both types of basic design problems encourage embedding and recursion but the emphases are different. One may argue that while additive processes run the risk of limiting one to discrete parts and preset structures, subtractive processes may encourage part embedding. The ice-ray examples [Stiny 1977] are comparable to the second type. Parameterized divisions provide flexibility in changing how the rule is applied similar to the subtractive approach. Rules are applied in changing ways. Özkar [2004] had argued previously that working with an interpretable whole provides for more creative operations than working with conceptually constrained initial shapes. Dividing the whole into consequent parts involves recursion as much as an additive process. But, since each recurrence deals with what is emergently left over from, or has been previously an embedded part of the step before it (rather than a composed new whole as in addition), these subtractive processes might be further explored as a way to convey to the students the emergence of new wholes alongside of recursion.

Oxman [2002] describes emergence as visual cognition and distinguishes between the syntactic and semantic structures involved. The advantage of shape grammars over the symbolic rule-based approaches has been the open door for emergence, enabled by seeing. Semantics are left to the subject and syntax is left to circumstance. Emergence of new shapes in additive shape computations has allowed for dynamic and creative generative processes with indeterminable results and large design spaces. Moreover, the embedding part relation allows for emergence of new parts all the time.

The two examples in the next section follow along the thread of additive schemas and emerging shapes. One of them starts the students with a given additive schema and the other starts the students with an already designed whole asking for what possible schemas may have been used. While inquiring into how formal ways to talk about design can be introduced to first-year architecture students, this paper draws attention to expansions of these notions with visual evidence from these two basic design examples.

In line with Stiny's conceptualization, a design schema is also described as a visual form of generic design knowledge [Oxman 2000]. In the context of design education, Oxman and Streich [2001] propose that the acquisition of the ability to represent design knowledge and the basic visual schema are part of learning in design. With emphasis on the cognitive aspects of what they imply, a computer model has been developed to support the representation of visual schemas and cognitive processes in design [Oxman 2000]. Additionally, with the purpose of providing a formal description of design exploration with sketches, Prats et al. [2009] show how schemas correspond to design rules in a process and utilize visual schemas to correlate different designers' visual rules. However, the existing literature does not discuss visual schemas with a particular reference to foundational design education.

The discussions below show some schemas and some visual rules they encompass, with illustrations from student works. The works are by Havva Elif Koç in the first exercise, and by İlker Teker and Sabiha Göloğlu consecutively in the second exercise. All three were students in the first-year design studio at the Middle East Technical University Department of Architecture between the years 2004 and 2006.

Before looking at the examples, here is the list of visual schemas that will be discussed:

(Schema 1) $x \rightarrow \Sigma t(x)$

(Schema 2) $x \rightarrow x + t(x)$

(Schema 3) $x \rightarrow x$

(Schema 4) $x \rightarrow x'$

(Schema 5) $x \rightarrow x + x'$

These all have their place in Stiny's specification of rule types [2009: 168] and can be extended to include $b(x)$, $b(t(x))$, $prt(x)$, and $b(prt(x))$ as will be discussed in the examples. Visual rules that are depicted work within these schemas. This paper is limited to only a few of these rules. In the studio environment, the number of rules to be discussed increase and the meticulous representation of each and every one of them loses any practicality. The argument based on these cases is that when different students operate within one schema, rule variations are good to compare and learn from. There are too many visual rules and no grammar that can be finally set whereas schemas are general devices that help compare, group, categorize, associate, contextualize and vary rules. Students do not discuss every single possible rule, but do have the chance to observe and discuss the fact that rules of the same schema are good for doing different things.

Practicing with an additive visual schema

The first example is based on a recursive prescription given by the instructors. This exercise makes it possible to concentrate on a small number of rules and schemas, and introduce the notion of material properties. Using the given visual schema, the student looks for and works with changing form relations of parts and wholes. New parts and wholes as well as other visual schemas constantly emerge. Shapes, weights and labels come into play. The task turns out not to be simply the compositional one prescribed in the beginning. Here, the student is encouraged to see, within the bounds of a given set of confining rules, and to explore multiple variations. The analysis shows the visual exploration that is possible with an identified formal schema.

In this first example, an additive schema is verbally expressed to the students at the start. Students are expected to work with primitives that they defined (independent from this new problem) in a previous exercise. These primitives, however, consist of parts that they can modify in terms of weights. In the end, the sum of these primitives yield new wholes (more than the sum of parts) because their parts can be configured into new shapes. The task in the assignment is to put together nine square units in a square grid, with the aim of creating a composition. Composition, in the context of the studio, implies a unified whole of parts/elements. The unit is itself a two-dimensional composition done in the previous assignment. Students are allowed to modify some aspects of the composition as necessary and according to how they adapt it to its new environment. Extreme changes are not encouraged.

The end result of one student's work is presented below in fig. 1.

Fig. 1.

When starting this particular composition, the student already had at her disposition a unit shape from a previous assignment. Let us for now only take the boundaries of the maximal planes ignoring the different tones/weights (fig. 2).

Fig. 2.

The general schema for the exercise

The student is asked to put together nine of these units to form a square. Euclidean transformations of the unit are allowed. The description of the assignment can be illustrated by the schema x → Σt(x) **(Schema 1)**. A visual rule that complies with this schema and describes the problem well is **Rule 1**. One square is replaced with nine, each some Euclidean transformation of the first.

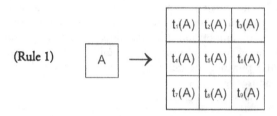

(Rule 1)

t₁(A)	t₂(A)	t₃(A)
t₄(A)	t₅(A)	t₆(A)
t₇(A)	t₈(A)	t₉(A)

A →

The schema for experimenting with couplets of units

As the task is not necessarily a one-step computation, the student also may choose to use **Schema 2**, x → x+t(x), to start putting units together two at a time. Many rules fit under this schema, one of which is "add one square to any one side of the original square." The student could have had an alternate spatial relation/rule where the two adjacent units have line parts that align as in **Rule 3** rather than meeting perpendicularly as in **Rule 2**.

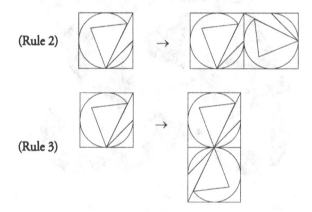

(Rule 2)

(Rule 3)

Alternatively the student could have chosen from among many other rules of the kind x → Σt(x) **(Schema 1)** if she wished to try more of the parts together – for example, groups of three or more – at one time.

The schema for seeing new relations between elements

In a further step, the spatial relation formed as a result of **Schema 2** above evokes the following **Rule 4**. In short, the student sees a *corner*. The schema to acknowledge this is the act of seeing x → x **(Schema 3)**. **Rule 4** serves as the basis for a new rule **(Rule 5)** within the previous schema x → Σt(x) **(Schema 1)** where the student more or less clarifies the aim to create a new shape with parts of several units.

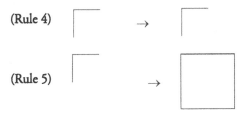

(Rule 4)   →

(Rule 5)   →

She identifies a shape x within what is given and also establishes that $\Sigma t(x)$ can be a desired shape (in this case a square). This new $\Sigma t(x)$, and its complementary parts can also be produced in other ways, for example with **Rule 6** again within the same schema.

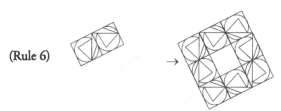

(Rule 6)   →

By still keeping in mind the rest of the shape while forming a square with the four corners, **Rule 5** is semantically guiding **Rule 6**. In this case, x and prt(x) are considered simultaneously and dependently for the creation of a new shape: prt(x) → $\Sigma t(prt(x))$ is a condition to x→$\Sigma t(x)$. One sees and operates on a shape that is a part of another shape, and the operation changes the whole shape at the end. The reason for the student to try out **Rule 6** is because she knows of and wants to apply **Rule 5**.

Simplified version to show the rich compositional variety

To retrace the same computation and to show the abundant variations the student may have or could have tried, let us exclude some of the features to emphasize what is going on with this emergent shape and new schema. The updated and simplified version of **Rule 1** is **Rule 7** from **Schema 1** x → $\Sigma t(x)$.

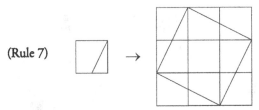

(Rule 7)   →

The right side of this rule shows all of the 8 transformations of the initial unit with the symmetry group of 8 (fig. 3).

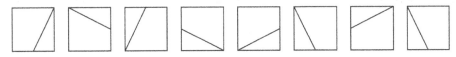

Fig. 3.

Rule 2 can be applied to put two units together in 64 different ways using the symmetries. The outcomes are reduced to 22 after similar ones are omitted (fig. 4).

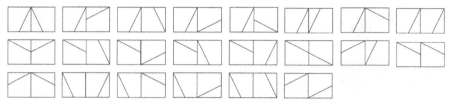

Fig. 4.

Including weights

There are already 22 ways to go at the very beginning of trying out couplets. Moreover, the problem asks the students to work with different tones/weights. If black is assigned to the triangular area and white to the quadrilateral, the 22 pairs appear as shown in fig. 5.

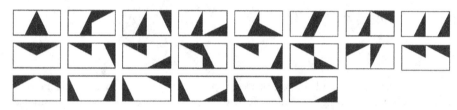

Fig. 5.

Following **Schema 2** $x \rightarrow x + t(x)$ again with some of these pairs, **Rule 8** can produce so many various results.

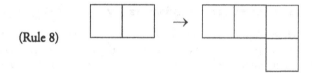

(Rule 8)

The student can come up with designs like those shown in fig. 6:

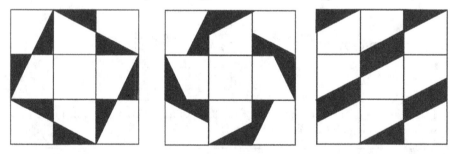

Fig. 6

When, however, the tones interchange between the triangular area and the quadrilateral, the number of alternating pairs rises to 85 (fig. 7) and eventual combinations of nine units multiply. The additive schemas guide the students to try out relations in similar ways but at the same time have them focus on seeing new relations among parts as options multiply due to different transformations and weights.

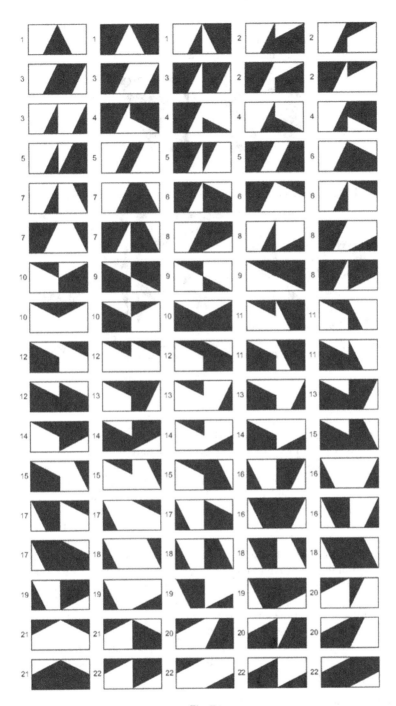

Fig. 7.

In this particular exercise, the student has more areas to assign tones to and more tones to assign than our analysis shows. Moreover she tries different combinations. Figure 8 shows two combinations she has in her final design.

Fig. 8.

Some possible outcomes just with changing the tones could have been those in fig. 9.

Fig. 9.

Overview

This exercise and its analysis demonstrate a few points. Firstly, however simple the shapes are, the students are dealing with large design spaces with multiple relations of parts and wholes at one time. The analysis given here only shows a small portion of it. Any production and discussion in the studio will also be limited to a small portion. Whereas the student is encouraged to enumerate as many possible options as possible, it is not practical and meaningful to talk of ALL the rules and relations as enumeration merely shows the abundance of choices with little clue to how the student is to guide himself or herself through these choices. Secondly, while the number of rules increases quickly, schemas recur, and talking in terms of schemas can be the students' guide to making meaningful choices. The analysis shows the generic schemas used. **Schema 1** $(x \rightarrow \Sigma t(x))$, given in the definition of the exercise, and **Schema 2** $(x \rightarrow x + t(x))$ are additive and promote recursion as a method of pursuing the problem. Within the recursions, students can exercise variation of relations and outcomes. That the rules vary from moment to moment within the same schema might be useful in terms of acknowledging different cognitive steps. **Schema 3** $(x \rightarrow x)$ is significant as it indicates moments of discovery or seeing. All three are included in the specification of rule types Stiny presented during his course lecture "Recusion, Identity, Embedding" given at SIGGRAPH [Stiny 2009].

Practicing recognizing various visual schemas

The second example illustrates a reversed relationship between recurrence (with an additive schema) and seeing. The task is reading a fellow student's design work to identify visually recursive patterns, relations, and visual schemas, and to improve on them. This time, it is the student who identifies some rules or schemas within an existing composition to follow and apply again. Seeing and doing are reversed. Whether the second one is an improvement over the first is not an issue. The example is more for comparison and an opportunity to talk about how to read design rules similar to Habraken and Gross's "silent game" [1987].

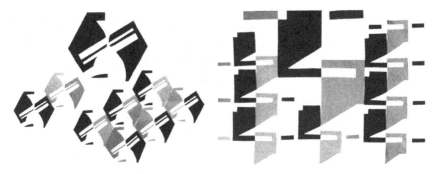

Fig. 10.

The general schema for the exercise

To start with, the general schema from the first student's work to the second's (fig. 10) is x → x′ **(Schema 4)** specified for general transformations and parametric variation. Shapes change a little; some parts and relations are maintained. The number of shape groups increases in the second work. There is still one big group more or less centrally located in both. The schema for both the first and the second work individually is x→Σt(x) **(Schema 2)**. The only exceptions in the second one are a few shapes in the top row where a few rectangular pieces, or t(prt(x))'s, come into view.

The schema for forming groups of elements

In the second work, the schema to acquire the module that consists of two parts seemingly in rotational symmetry with one another is x → x + x′ **(Schema 5)** and is visualized in **Rule 9.**

(Rule 9)

If instead the schema had been a more specific schema x → x + t(x) **(Schema 2)**, the composition shown in fig. 11 would have resulted.

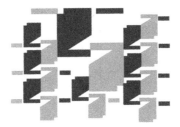

Fig. 11.

The schema for organizing the layout of the composition

Another schema recognized in the first work is the schema to organize the page. Parts are arranged in alignments and this could be specified in **Rules 10 and 11**. **Rule 10** is employed by itself in the first composition whereas the second student creates a variation of **Rule 10**, namely **Rule 11**, in the same schema and uses both. This is similar to the visual rules discussed in the context of other basic design problems [Özkar 2005b]. It is promising that not only schemas but sometimes rules as general as these are recurrent in various design problems. The two rules are especially significant to showcase the slight variations within a schema and what different outcomes can result. Here, the second student acknowledges the spatial relations, and reapplies them with slight variation with the aim of improving the overall design and how it fits an orthogonal page frame.

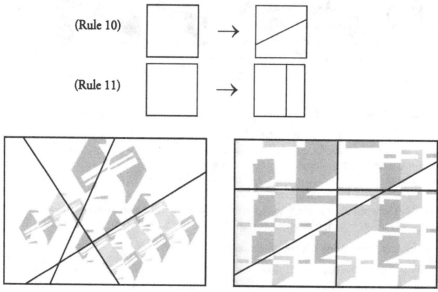

Fig. 12.

As a result of this variation, the axes for shape alignments differ from the first to the second. The second application of the schema takes into consideration the orthogonal frame of the canvas along with the angles of the parts.

The schema for aligning elements of the composition

Furthermore, if we look at what is aligned, we see that the second student reads in the first work the alignment of boundaries of parts b(prt(x)) and takes care to follow the same criterion. **Rule 12** from **Schema 2**, $x \rightarrow x + t(x)$, can be used to specify this, and the distance in between the two aligned parts can be varied to fit many possibilities.

(Rule 12)

Continuing to look at aligned parts, we see that since the alignment is of the boundaries and not of the whole shapes, there is variation in which side of the shape they are aligned with. **Rules 13 and 14,** again from the same schema as above, depict these relations.

Rule 13 aligns the edges of planar shapes on the same side, and **Rule 14** aligns them on opposite sides.

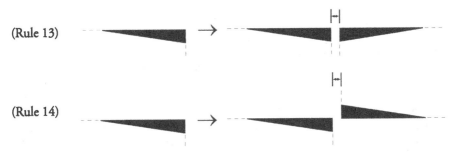

Since in most of the alignments one piece seems to be matching the void in the opposing piece as in **Rule 15**, these planes can also be interpreted as weighted lines. A thick line with assigned W_1 (black) is aligned with a thick line assigned W_2 (white) in **Rule 16**. In both students' works, as grey and black shapes align, white shapes emerge as parts to be picked out in the same schema. They are just as good to align with as the darker colored shapes.

Overview

There is simply no unique way to look at these compositions. The students read many rules which all fit into only a few schemas. Schemas are general and sharable. Rules are more specific, sometimes comprising weights, labels, or parametric values, and are hence greater in number. Mostly they are alternatives for one another in alternative computations of the same problem. For example Rules 12-16 above are all valid, even towards the same result, but open up different potentials. Also, rules are mostly coincident and may serve similar purposes. Weights can be specified in one way, but the end result of the rule could immediately invoke another way. A rule could be applied to align the black line with the grey line, only to find that the grey line is defined by its neighboring white line so that the black line is also aligned with the white line. If one part aligns, some other parts automatically do as well, and at different levels parts act together to form numerous wholes which correlate in the bigger whole. These changing hierarchies are not very easy for the students to grasp; to unite them in one schema would help. The important lesson here is that rules are part of a shared schema, which in turn helps the student talk about the rules and their differences.

To illustrate the same richness here is one other observation. The detailed nooks in the first student's work are simplified into singular nooks in the second's. This shows that schemas allow for the use of rules with different specificities. Parts of shapes can be ignored, proportions can be altered.

In this second example, we are looking at the process of the second student rather than the process behind the first composition. The analysis is based on a comparison of the two. It is a good example for showing the schemas identified by the second student. Weights, scale transformations and number of parts change but schemas are common. How rules within the same schema are applied or detailed show us variations of the same theme. Going back and forth between a schema and shape rules helps the student to talk about variations. The second student identifies some rules, speculating on what the first student might have been thinking about. They can be combined or modified, for example simplified, or there are simply different ways to talk about the same thing. There are many options, but schemas generalize them. The second student does not have to understand fully how the first student used a particular rule, but the schema is shared among the two.

There are more opportunities for the student to see different parts to the shapes in this second example because the primitives are not given, but identified in different ways again and again on the spot. She excludes details where she sees fit. She utilizes schemas of seeing: $prt(x) \rightarrow prt(x)$, $t(prt(x)) \rightarrow t(prt(x))$, and $b(prt(x)) \rightarrow b(prt(x))$. This is the kind of emergence mentioned in the beginning of this paper. Surprises occur in the additive processes, in this case mostly within **Schema 5**, $x \rightarrow x + x'$, because the wholes are more than the sum of parts. Parameters are not defined and the grammar is not as precise as in the ice-rays, but certainly those can come in more advanced exercises to follow.

Generic schemas work for discussing recurrence, emergence and variation

Formalisms are not a direct way of talking about design in the foundations studio where seeing and doing are emphasized. Formal ways to talk about design actions are conventionally encouraged and so far they have mostly been drawings, or the products themselves in addition to words that are transferred from daily life. A novice designer's vocabulary slowly develops through discussions. It is a worthy idea to include shape grammars to these formal ways, but shape grammars and proper visual rules are external devices that are difficult to incorporate into the already heavy intellectual and physical load at hand. Students are not going to be bothered with abstractions (the notion of computing with shape rules might as well be abstract to them) in an environment where they are constantly being encouraged to think visually and spatially, and the number of potential rules is quite high. Pointing at some relation while trying to come up with a good descriptive word, or at best a concept, is the practice. Nevertheless, rules can be talked of as simple as two shapes connected with an arrow, and to compare and contrast these, visual schemas seem to be relevant. Rather than a seamless mathematics of shapes that do not add up to a sensible grammar, the schema is easier to grasp, and it addresses, with benefits, the need for formalism.

The schemas discussed in relation to the two examples also signify the design knowledge conveyable in the foundations studios. Firstly, emergence is not only the occurrence of a new shape in a linear generative process, but also a recognition of a shape that was there before but not seen. This is generally what is referred to as the ambiguity in shapes, but can be extended to emergence if one understands it with reference to the changing context. In the examples discussed, shapes are parts of wholes and co-existing parts act in the emergence of new shapes despite the fact that these shapes are already there to begin with. Hence, there is a need to talk about emergence as ambiguity in a general sense. More generic schemas can be devised in future studies, for example to include b(x) or prt(x), as it is mostly boundaries or parts that the eye sees in a different light.

Similarly, recursion in design is more complex than a linear iteration. In basic design exercises, repetition or rhythm comes forth as an organizing schema and is rarely linear. It develops in multiple dimensions and multiple instances of it coexist. In architecture education, repetition comes forth as a strong schema for making sensible and unified wholes. The two exercises discussed above are good illustrations for rhythm in two dimensions that is more than merely a linear recursion.

Recursion and emergence culminate in variation, which includes not only repetition but also similarity and change all at once. A schema of variation could be:

Rule A $(x \rightarrow x + t(x))$ varies into **Rule B** $(prt(x) \rightarrow x + t(prt(x)))$.

Variation is a powerful tool for teaching design and is easily talked about with visual rules/schemas. For example, the schema $x \rightarrow x + y$ might help in comparing and contrasting rules where y is varied with one constraint and in understanding, recognizing, discussing organizational decisions. Visual rules help not only in generating alternative solutions but also in creating controlled variations of one solution as a way of trying and erring and retracing. Being able to vary (according to a theme if needed) seems to be an important skill to acquire for beginning design student.

In short, schemas help identify, generalize and convey the key aspects of relational and reflective thinking: recursion, seeing emergent shapes as well as parts, boundaries, and relations, and building up variation, prior to learning proper shape rules and design grammars.

This paper can be taken as a preliminary study to place formalism in the studio. More inquiries are required to develop and test ways to incorporate visual schemas to design studio talk systematically. Interdisciplinary collaborations may come in handy for addressing pedagogical, cognitive and instrumental aspects of the problem in a unified manner. Moreover, there are many more rule types to be explored in conjunction with basic design exercises. Hierarchic relations to be explored between the types may also be helpful for students.

These analyses set up examples for better understanding design computations with visual schemas. They show that visual computations coexist at multiple levels in visual design processes, that parts and wholes are interrelated, and that different wholes are causally related. Both examples show that an intensity of different simultaneous perceptions of parts exists. There is a complexity due to the parallel processing of shapes. Schemas are a way for the beginning students to understand this complexity and how to be active within it. Moreover, visual computations are not just literally limited to the visual. They are coupled with labels (that is, information relevant to the application of the rule), such as assumptions about a certain part (e.g., the Gestalt notion of continuity) and material properties (e.g., properties that weights are assigned to). Schemas encompass all these traits and set up a good prequel to learning about more explicit shape computations in more advanced level design studios.

Acknowledgments

The basic design works that are analyzed are outcomes of student exercises given in the Basic Design Studio at the Middle East Technical University, Department of Architecture in 2004 and 2005. The course at the time was taught by Selahattin Önür, Tuğyan Aytaç Dural, Nihal Bursa, Erkan Gencol and Mine Özkar. The exercises were designed by the entire team. The shape analyses, rules and schemas that are the subject matter in this research are all produced retrospectively by the author. The author would also like to thank George Stiny for countless valuable discussions on visual schemas.

Bibliography

BROADBENT, Geoffrey. 1979. The Development of Design Methods. *Design Methods and Theories* **13**, 1: 25-38.

BROWN, David. 2003. Learner-Centered Conditions that Ensure Students' Success in Learning. *Education* **124**, 1: 99-104, 107.

CELANI, Maria Gabriela Caffarena. 2002. CAD – The Creative Side: An Educational Experiment that Aims at Changing Students' Attitude in the Use of Computer-Aided Design. Pp. 218-221 in *SIGraDi* 2002 (Proceedings of the 6th Iberoamerican Congress of Digital Graphics, Caracas, Venezuela, 27-29 november 2002). http://cumincades.scix.net/data/works/att/7520.content.pdf (last accessed 04 November 2010).

CHASE, Scott C. 2005. Generative design tools for novice designers: Issues for selection. *Automation in Construction* **14**, 6: 689-698.

DO, Ellen Yi-Luen and Mark GROSS. 2001. Thinking with Diagrams in Architectural Design. *Artificial Intelligence Review* **15**, 1-2: 135-149.

FINDELI, Alain. 2001. Rethinking Design Education for the 21st Century: Theoretical, Methodological, and Ethical Discussion. *Design Issues* **17**, 1: 5-17.

HABRAKEN, H. John and Mark GROSS. 1987. *Concept Design Games* (Books 1 and 2). Cambridge, MA: Design Methodology Program.

KIRSCH, J. L. and R. A. KIRSCH. 1986. The Structure of Paintings: formal grammar and design. *Environment and Planning B: Planning and Design* **13**: 163-176.

KNIGHT, Terry. 2000. Shape Grammars in Education and Practice: History and Prospects. *International Journal of Design Computing* **2** (1999). Available at: http://www.mit.edu/~tknight/IJDC/ (last accessed 04 November 2010).

OXMAN, Rivka and Bernd STREICH. 2001. Digital Media and Design Didactics in Visual Cognition. *Architectural Information Management* (Proceedings of the 19th eCAADe Conference, Helsinki, Finland, 29-31 August 2001), pp. 186-191.

OXMAN, Rivka. 2008. Digital architecture as challenge for pedagogy: theory, knowledge, models and medium. *Design Studies* **29**, 2: 99-120.

———. 2002. The thinking eye: visual re-cognition in design emergence. *Design Studies* **23**, 2: 135-164.

———. 2000. Design media for the cognitive designer. *Automation in Construction* **9**, 4: 337-346.

ÖZKAR, Mine. 2004. Cognitive Analyses and Creative Operations. Pp. 219-229 in *Visual and Spatial Reasoning in Design III*, John Gero, Barbara Tversky, Terry Knight, eds. Key Center for Design Computing and Cognition, University of Sydney, Australia.

———. 2005a. Form Relations in Analyses by Denman Waldo Ross: an Early Modernist Approach in Architectural Education. Pp. 322-328 in *Aesthetics and Architectural Composition* (Proceedings of the Dresden International Symposium of Architecture 2004), Ralf Weber and Matthias Amann, eds. Mammendorf: pro Literatur Verlag.

———. 2005b. Lesson 1 in Design Computing Does not Have to be with Computers: Basic design exercises, exercises in visual computing. Pp. 311-318 in *Digital Design: The Quest for New Paradigms* (Proceedings of the 23rd eCAADe Conference, Lisbon, Portugal, 21-24

September 2005), José Pinto Duarte, Gonçalo Ducla-Soares, and Zita Sampaio, eds. Lisbon: eCAADe and IST.

PRATS, Miquel, Sungwoo LIM, Iestyn JOWERS, Steve W. GARNER and Scott CHASE. 2009. Transforming shape in design: observations from studies of sketching. *Design Studies* **30**, 5: 503-520.

SASS, Larry. 2005. A wood frame grammar: A generative system for digital fabrication. *International Journal of Architectural Computing* **4**, 1: 51-67.

STINY, George and James GIPS. 1972. Shape Grammars and the Generative Specification. Pp. 125-135 in *Best computer papers of 1971*, O. R. Petrocelli, ed. Philadelphia: Auerbach.

STINY, George. 1977. Ice-ray: a note on the generation of Chinese lattice designs. *Environment and Planning B: Planning and Design* **4**, 1: 89-98.

————.1985. Computing with form and meaning in architecture. *Journal of Architectural Education* **39**, 1: 7-19.

————. 2006. *Shape: Talking about Seeing and Doing*. Cambridge, MA: MIT Press.

————. 2009. Shape grammars, Part II. Pp. 99-172 in *ACM SIGGRAPH 2009 Courses*, Article 22. New York: ACM. http://portal.acm.org/citation.cfm?doid=1667239.1667261 (last accessed 16 November 2010).

About the author

Mine Özkar, an architect by training, holds a Master of Science degree in Architecture Studies and a Ph.D. in design and computation from the Massachusetts Institute of Technology. She has been coordinating and teaching undergraduate and graduate level design studios and computational design courses in the departments of architecture at the Middle East Technical University since 2004 and as an adjunct assistant professor at the Istanbul Technical University since 2009. Pedagogically, she focuses on the integration of computation to architectural design education starting from the first year studios. Some of her collaborative research is on information and computing technologies in design processes, visual computation, and computer implementation of part relations of shapes. She has also published on the history, theory, and practice of foundational design education and its integration with computing, as well as on developing methodology for curriculum changes in architectural education as part of a nationally funded research project.

M. Piedade Ferreira

Curso de Doutoramento em
Arquitectura
Faculdade de Arquitectura
Universidade Técnica de Lisboa
Rua Sá Nogueira
Pólo Universitário, Alto da Ajuda
1349-055, Lisbon PORTUGAL
mpferreira@fa.utl.pt

Duarte Cabral de Mello*

*Corresponding author

Faculdade de Arquitectura
Universidade Técnica de Lisboa
Rua Sá Nogueira
Pólo Universitário, Alto da Ajuda
1349-055, Lisbon PORTUGAL
dcm@fa.utl.pt

José Pinto Duarte

Faculdade de Arquitectura
Universidade Técnica de Lisboa
Rua Sá Nogueira
Pólo Universitário, Alto da Ajuda
1349-055, Lisbon PORTUGAL
jduarte@fa.utl.pt

Keywords: shape grammars; corporeal
architecture; choreography;
performance; attunement

Research

The Grammar of Movement: A Step Towards a Corporeal Architecture

Presented as a poster at Nexus 2010: Relationships Between Architecture and Mathematics, Porto, 13-15 June 2010.

Abstract. This research uses shape grammars as the basis of a computational tool to explore the relationship between the human body in motion and space, aiming to develop further knowledge about cognition and architecture. Artistic and scientific tools and methods already used to develop these concepts are being studied in order to create a new tool that will help us to understand, through simulation, how the body mediated through architecture can influence human cognitive response and thus behaviour. The goal is a methodology for the design of a "corporeal architecture" that can create a naturally immersive environment in which the ability of its geometry and physical properties to conduct or induce body movements in space for specific purposes can generate experience. Also discussed is the potential of the tool proposed for the study of the human body in movement as a generative strategy in architecture. In describing the parameters and criteria chosen to develop our software, we exemplify briefly how a shape grammar, as a system of rules, can be used to generate sequences of actions, establishing the idea that human behaviour in space can be composed as choreography and provide a means of considering architectural space not only in terms of shape but particularly in terms of life.

The human body, existential space and (virtual) reality

Cyberspace is a very relevant aspect of human interaction today, both for leisure and professional purposes. Most people use email and web-based communication or the new social networks to contact friends, family, partners, clients or employees. These media have also begun to be used for political campaigns and other important areas of public life due to their powers of persuasion and immersion in the minds of users. Human-machine interfaces, such as the Nintendo Wii™ video game console which allows the whole body to move, are developing rapidly but, in general, most of them require repetitive, mechanical and passive body postures, over-stressing the body as a whole. The result is that many people are starting to suffer from health problems such as obesity, poor eyesight, muscular and skeletal disorders or even insomnia and depression.

The problem of the mechanisation of the human body and the loss of individuality was already being discussed in the arts at the beginning of the last century. One example

Nexus Network Journal 13 (2011) 131–149
DOI 10.1007/s00004-011-0058-4; *published online* 26 February 2011

is Charlie Chaplin's *Modern Times* which, with characteristic humour, warns against the mechanisation and systematisation of life, and extremely repetitive tasks in labour and leisure activities and even for survival (fig. 1). The film's main character, the "Little Tramp" is shown in a series of comical but paradoxically tragic sketches in which technology is seen as so aggressive and oppressive that he cannot cope with it and ends up having a nervous breakdown, losing his job and wandering homeless in the streets.

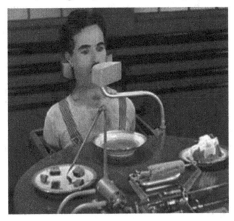

Fig. 1. Worker feeding machine. Still from Charlie Chaplin's *Modern Times* (1936)

Given the historical and ethical fact that the timeless role of architecture is to create metaphors of human existence as a way of building and supporting ourselves in the world, what can be concluded, with regard to architecture as art and technology, from the growing activity in cyberspace which forms part of contemporary social interaction? Can this growing activity in virtual worlds be a true indicator of architecture's failure as an immersive system, leading to the search for other experiences in parallel universes as a result of collective dissatisfaction with the built world? For Juhani Pallasmaa,

> The dehumanization of contemporary architecture and cities can be understood as a consequence of a neglect of the body and the senses [...] an imbalance in our sensory system, [with] today's growing experiences of alienation and loneliness related to a certain kind of pathology of the senses [Pallasmaa 2005: 16-19].

Pallasmaa emphasises this beyond architecture by stating that contemporary culture is heading towards a terrifying de-sensualisation and de-eroticisation of human relationships in reality. Thus, virtual reality interaction may be demanding a paradigm shift in architecture, forcing it to reconnect with the human being and human emotions by providing stimulating interaction with the physical environment, including built forms, other humans and animals. A "corporeal architecture" that aims to connect with the dweller by triggering his senses, would stimulate his mind and body holistically and provide a naturally immersive environment, offering balance and a sense of well being.

Steps towards a corporeal architecture

> *To think about architecture is to think in another way [...] is to accept chaos* [and] *live it with the feet*
>
> Bruno Queysanne [1987: 95-98]

One of our approaches to this problem described was to try to understand what kind of means and what type of knowledge of the human body the new media uses to enable it to absorb people for so long. Recent technological progress has allowed many sciences related to the new media to develop very fast, due to their use of computation, programming and artificial intelligence, namely tools that allow complex ecologies to be understood and reproduced through algorithms. These have influenced the latest developments in mind and brain studies, such as cognitive psychology, the physiology of the human body (mind) and the latest discoveries in the neurosciences, such as emotional intelligence, mirror neurons theory and the "embodied mind" paradigm [Silvério Marques 1990: 159-187].

In architecture, the use of these tools could open up the path proposed by Christopher Alexander in the article "A City is not a tree" [1965], in the sense that they could allow for the "artificial" reproduction and simulation of what he calls "natural" cities. Some strategies already use new media to study behavioural phenomena on an urban scale by means of simulation and cross referencing large amounts of data. One example is Space Syntax, based on Hillier's "The Social Logic of Space" [1996]. The application of a generational model with this purpose can be very useful since it offers an understanding of the essential conditions that ensure efficient architecture in terms of human behaviour, Alexander's "natural" approach.

Bearing these concepts in mind, and working within a broad frame of reference that includes the philosophy of action, psychology, anatomy, phenomenology, semiotics, shape grammars and media studies, we have been constructing a body of knowledge on the perception of the human body, which is still disaggregated, and are trying to understand how architecture can incorporate this in order to become more appealing to the senses and have a greater impact on memory. This could involve subjective experience and also appeal to collective memory, whilst reflecting natural human gestures to protect the body, the source of architecture's archetypes. We call such architecture, "a corporeal architecture".

Hopefully, a corporeal architecture will reflect the elementary characteristics that have allowed human beings to group together to form social bonds and cities – what Cabral de Mello [2007] calls "architectural deep structure or genotype",[1] namely the mechanisms underlying the creation of artefacts, tools and architecture, the most complex example, as an essential extension of the human body (fig. 2).

Fig. 2. Still from Stanley Kubrick's *2001: A Space Odyssey* (1969). A step in human evolution, the invention of the tool, using body movement

Some archetypical traces of a grammar of movements

To Vitruvius, the human body was the source of all geometry. As a man of his time, Vitruvius followed the Pythagorean tradition and considered nature the source of all knowledge, transcribed in the form of the perfect number, ten. To Vitruvius, the human body is nature's prime work, incorporating the only creature with ten fingers, and the ability to think, create and manufacture objects. The Vitruvian man does not move to produce geometry, "flat on his back and passive, he is *made* to produce it. The active agents, as Vitruvius tells it, are the compass and the set square. At once a metaphysical proposition and a ritual formula, Vitruvian man is also and above all the architect's template" [McEwen 2003: 181]. So, the Vitruvian man can be made to produce a circle, and a sphere, since to the Stoics, in the words of Cicero, as quoted by McEwen, they alone "possess the property of absolute uniformity in all their parts", "no other shape can maintain the uniform motion and regular disposition of heavenly bodies" [Cicero, *De natura deorum* 2.47, 2.48 quoted in McEwen 2003: 160].

Being static in itself, the "architect's template" is a tool that represents the canon of human body proportions, and its use generates motion in space. The results produced are meaningful through the action of the architect and the culture to which he belongs.

Leonardo da Vinci rediscovered the Vitruvian man in the Renaissance and created its most famous descriptive drawing in which Man is not static at all, but standing and in motion. Leonardo called this system a Canon of Proportions in which sixteen possible positions for the human body are systematized in the same drawing, generated by variations on two main positions overlapped and inscribed within the cosmological symbols of the circle and the set square. It must be no coincidence either that Leonardo's Canon of Proportions (fig. 3) depicts sixteen actions. As quoted by McEwen [2003: 50-51], Vitruvius says in *De arquitectura* that "the Romans recognized both 6 and 10 as perfect numbers and combined them to make the supremely perfect number 16, finding the rationale for this in the foot, which had sixteen fingers", thus 16, the number 4 squared, was in McEwen's words "the agent and evidence of Roman order and also represented the Etruscan division of the sky into four (two squared) cardinal signs [2003: 51].

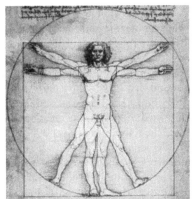

Fig. 3. The Canon of Proportions, Leonardo da Vinci's depiction of the Vitruvian Man, c. 1487. Gallerie dell'Accademia, Venice

Leonardo's interpretation of the Vitruvian man is standing, following the Roman theology of the feet. According to McEwen,

> Arrian, writing in the second century A.D., tells the following story about the arrival of Alexander the Great in India:
> *Some Indian sophists, the story goes, were found by Alexander in the open air in a meadow, where they used to have their disputations; when they saw Alexander and his army, they did nothing than beat with their feet on the ground they stood on. When Alexander enquired through interpreters what their action meant, they replied: "King Alexander, each man possesses no more of this earth than the patch we stand on"* [McEwen 2003: 52].

This story gives some meaning to the similarities found between both Vitruvius's description, Leonardo's representation of the cosmological Man and the Hindu archetypical depiction of Shiva as Nataraja, the Lord of Dancers. The visual image of Nataraja achieved canonical form in the bronzes cast during the Chola dynasty in the tenth century A.D. and are often said to be the supreme statement of Hindu art (fig. 4).

Fig. 4. Shiva as Nataraja, canonic depiction from the tenth century A.D.

Shiva as Nataraja is the arch-yogi of the gods and the system is a representation of the dance that generated yoga, which in Sanskrit means union of body, soul and cosmos. It depicts the four-armed body of Shiva, each arm representing the four cardinal directions and therefore, a square. Through its choreography, the balance between the motion of the limbs represents the whirling of time and the immobility of the serene expression on his face; the paradox between Eternity and Time. The ring of fire and light, the circle that circumscribes the body, represents the entire universe in cyclic rhythm and in union through motion. The gestures (*mudras*) and objects in each of Shiva's hands represent the beating pulse and the sound that makes Shiva dance, the first element of the universe and the most pervasive. In Hindu mythology this sound generated the first grammar of Sanskrit, which Shiva transmitted to Panini, the great Sanskrit grammarian, and the first verse of his grammar is called the Shiva Sutra [Goel 2001]. The 'grammar' of movements encoded in the Nataraja depiction of Shiva was transmitted from generation to generation, extending into many symbolic gestures and actions related to spirituality, dance and religion. These yoga postures provide a link to some of the most primitive traditions of reverence to nature, being based on elementary and symbolic geometrical shapes such as the triangle, square, pentagon and hexagon, and embedded in the matrix of our collective subconscious. In eastern culture, Yoga is believed to have a very powerful psychophysiological effect on the individual, expanding his consciousness and improving his overall health and longevity.

The human body in motion as geometrical source and spatial generator

The geometrical proposition of the human body had much focus in the German and the Russian avant-gardes of the twenties, especially at the Bauhaus, where life drawing had always been part of the school's curriculum. Most Bauhaus' Meisters such as Feininger, Itten, Klee, Schlemmer, Kuhr, and later on, Joost Schmidt held this course successively, according to individual focuses. For example Itten concentrated his studies on expressiveness giving focus to the rhythmical coordination of the body limbs and the body structure as a whole, while Klee was interested in the representation of the tectonics of the human body through linear drawings, where the articulations necessary to movement were emphasized as dots.

Oskar Schlemmer, first hired as Master of Form at the Bauhaus theatre workshop in 1923, based his course on "Man" (1926/27) on the study of Dürer's system of proportion, Leonardo's Canon of Proportions and the golden section. He included in the course curriculum the study of human biology and chemistry, the stages of growth from gestation to maturity and also notions of psychology and philosophy. Schlemmer developed systems to study the mechanics and kinetics of the human body through notation and staged diagrams of movement in space (fig. 7).

 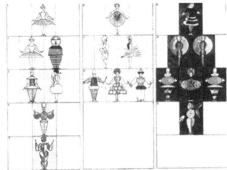

Fig. 5. Schlemmer and the Bauhaus's theatre group on the school's stage

Fig. 6. Schlemmer's designs for the Triadic Ballet, 1922 and 1926

These studies were the basis for his choreographies, culminating in the 'Triadic Ballet', his most famous piece (figs. 5 and 6). In his diary of July 5, 1926, Schlemmer explains why his ballet was given the name Triadic,

> Because three is a dominant number, in which the unitary 'self' and his opposed dualist are suppressed, starting the collective ... After that comes five, then seven and so on. The ballet should be understood as a dance of the triad, the switch from one with two, then three. ... Further, the triad is shape, colour, space; the three dimensions of space, height, width and depth; the fundamental shapes, sphere, cube and pyramid; the fundamental colours, red, blue and yellow. A triad of dance, costume and music [Schlemmer 1987: 88, translated from the Spanish version by the authors].

According to Pythagorean thinking, the number three was the number that embraced the totality of existence and mathematically originated the possibility of palpable extent. For Plato three was the number of the soul and for Aristotle, it was a beginning, a middle and an end [Le Corbusier 1954: 65-71].

To Schlemmer, space was the unifying element in architecture and the common denominator of the many interests amongst the Bauhaus staff. According to Goldberg, "what characterized the 1920's discussion on space was the notion of *Raumempfindung,* or "felt volume", and it was to this "sensation of space" that Schlemmer attributed the origins of each of his dance productions" [Goldberg 2006: 104] (fig. 8). Schlemmer's system was based on place geometry, coordinating simple elements such as the straight line, the diagonal, the circle and the curve. According to this theory, "a stereometry of space evolves, by the moving vertical line of the dancing figure" [Goldberg 2006: 104].

This stereometry could be 'felt' if space was imagined as being filled with a soft malleable substance in which the figures of the sequence of the dancer's movements hardened as negative form.

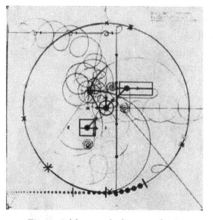

Fig. 7. Schlemmer's diagram for "Gesture Dance", 1926

Fig. 8. Oskar Schlemmer, drawing from *Mensch und Kunstfigur*, 1925

These abstract theories were illustrated in 1927 by Schlemmer and his students, at the Bauhaus stage in a dance called 'Dance in space (Delineation of Space with Figure)'. The square surface of the stage floor was divided into bisecting axes and diagonals, circumscribed by a circle and afterwards taut wire was run across the empty stage, defining the 'felt volume' diagram of the cubic stage space (fig. 9). A multiple exposure photograph of this performance by Lux Feininger gives us an image that resembles a living performance of a standing Vitruvian man in motion, generating volume with his body in the void of space (fig. 10). As a self-confessed admirer of classical philosophy, from which he drew support for his aesthetics and ethics according to the mythological opposition between Dionysus and Apollo, it is most likely that Schlemmer devised his "Mathematical Gesture Dance" using this phenomenological approach to Vitruvius's canonic tradition.

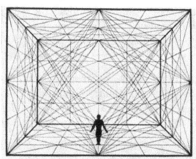

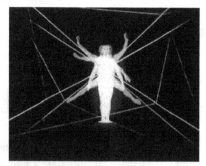

Fig. 9. Schlemmer's drawing for "Figure in Space with Plane Geometry and Spatial Delineations", 1927

Fig. 10. "Dance in Space (Delineation of Space with Figure)", multiple exposure photograph by Lux Feininger, Bauhaus Stage demonstration, 1927

Praise of the Pythagorean tradition continued with Le Corbusier and the Modulor (1948-1955), considering Mathematics "the majestic substructure conceived by Man to grant him comprehension of the universe" [Le Corbusier 1954: 71]. The Modulor (fig. 11) was devised as a measuring tool that systematizes the mathematical wisdom of the Pythagorean Triad and Duality and the Fibonacci sequence, and its creator hoped that it could relocate architecture in relation to the human scale, as it was based on the systems of the proportions of the human body, the Golden section and the Vitruvian man. To Le Corbusier, it would be a tuned measuring instrument which, combined with the technical resources of his time, "could make the good easy and the bad difficult" [Le Corbusier 1954: 58], facing the challenges and the growing complexity of the machine age.

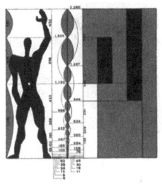

Fig. 11. "The Modulor", Le Corbusier, (1948-1955)

Le Corbusier wanted to express in his work the belief that "only the architect can strike the balance between man and his environment (man = psycho-physiology; his environment = the universe: nature and cosmos)" [1954: 111], a rather holistic view that also recalls humanist thinking. Paradoxically, being the creator of a system of rules that recalls many Renaissance authors such as Alberti, who considered the eye as the supreme organ of perception, Le Corbusier states in his final considerations in the Modulor, "I have stayed within the realm of concrete things, within the field of human psycho-physiology. I have concerned myself only with objects falling under the jurisdiction of the eye [1954: 184].

Le Corbusier's suspicion of the Renaissance architects was based on his understanding of their architecture as more a product of subjective and individual spiritual quests than a commitment to a social or universal philosophy. Ironically, his overall understanding of the "Universal Man" as an inflexible *tabula rasa* that scorned cultural, genre and emotional differences would be the cause of the Modulor's decline, when it became seen as a static, closed and abstract representation of Man. In Pallasmaa's words,

> The modernist idiom has not generally been able to penetrate the surface
> of popular taste and values, [it] has housed the intellect and the eye, but it
> has left the body and the other senses, as well as our memories,
> imagination and dreams, homeless [Pallasmaa 2005: 19].

Regardless of this, it should be noted that twenty-five years before designing the Modulor, in *L'Esprit Nouveau* Le Corbusier expressed the need for rules in architecture based both on scientific knowledge and art and experiment, a scientific aesthetics that combined reason and intuition. This would be based on his studies of cubist painting and sculpture and the musical concept of harmonics, applied to architecture in pursuit of the "fourth dimension, the moment of boundless freedom brought about by an exceptionally happy consonance of the plastic means applied in a work of art" [Le Corbusier 1954: 29-32] he key to aesthetic emotion being a function of architectural space. Even though the Modulor, as a system of rules, was not complex enough to deal with the challenges of its time, reducing them exclusively to a problem of proportion, its use was intended to open up the path for the architect to work with more assurance, letting intuition flow and making art easier. It can therefore be said with some assurance that if Le Corbusier had had at his disposal the computational means and knowledge available today, the Modulor would probably have produced very different results, encompassing many more parameters and criteria and opening up the path to a more individualized, subjective and "corporeal" architecture.

A New Tool: Genera(c)tiveHumanoidLifeForm: the simulation of simple movements to generate complex shapes / spatial relations with the dynamics of the human body

> We should start from the elemental. What does this mean? That we start
> from the plane, the line, the simple surface and that we start from the
> simple composition of surfaces, using the body

Oskar Schlemmer, diary entry, May 1929 [1987: 112; our translation]

After Le Corbusier's failed attempt to definitively fill the gap between the human body and architecture, there has been a great deal of criticism but little else has been done to solve the problem in an operative manner. In industrial design, especially in human machine interfaces or in prosthetics, there have been considerable developments that have led towards a kind of "body-tailored" design in which the fundamental concepts developed are flexibility, responsiveness and intuitive use, resulting in more intelligent and corporeal designs. This has been made possible due to recent developments in mind and brain theory and the new digital technologies that allow all kinds of information, including biomechanical processes such as movement, to be translated into a mathematical or programming language.

These tools have been extensively developed in cinema in the so-called computer generated images (CGI) or animations, in which characters are animated as digital puppets and the actor's job is to fill them with life, using his emotions, with all his physical expressions being transposed remotely to the puppet using motion capture hardware interfaces connected to his body in specific places. This information is interpreted by software that translates the information and generates algorithms to animate the puppet, character or avatar. Such algorithms are becoming increasingly complex and more accurate at an astonishing speed. Computer generated animations can mimic body language so faithfully that they can already establish the same kind of empathy or *attunement* that a good actor can create with his public, using his own body. Digital animation also allows the creation of all sorts of objects and geometries, through parametric design and the use of topological geometries, surpassing Euclidian limitations. This allows for the simulation and rendering of zero gravity and underwater environments and also different weather or lighting conditions.

In architectural design, digital modelling and rapid prototyping, tools are gradually replacing analogical design systems, but there is still no single tool that links the psychophysiological characteristics of dwellers and architectural space together as parameters or rules in the generation/simulation of personalised designs which are body-tailored, empathic and therefore "corporeal". Dweller performance and behaviour in space is usually the ultimate architectural test, only possible after construction and generally involving a great deal of uncertainty. Although the use of such tools will not, in themselves, allow for greater individuation of the user's in-space dwelling or generation, they can certainly help to increase the level of complexity and surpass the generic limitation of other systems such as the Modulor by encompassing rules and parameters that simulate complex bodies and situations.

This current investigation has been trying to develop such a tool and we have started to write an algebraic synthesis of the human body, hoping to transfer the biomechanical process of movement into digital information in a way similar to the method used in artificial life studies and robotics. We wish to open up the view of the canonic Vitruvius tradition by enlarging the universe of the human body types considered, which will allow for the simulation of a greater number of different actions and, consequently, spatial relationships and situations. In our program it is possible to choose different types of body, due to the careful introduction of the values that were chosen as parameters to be inserted in the programming code, resulting in: Male, Female, Child, OverAveragePerson and UnderAveragePerson (fig. 12).

In writing the body of the character in the programming language, the "Genera(c)tiveHumanoidLifeForm", it was necessary to find a way to draw a synthesis of simple geometric shapes that could be used to simplify a human body, bearing in mind that the model drawn must offer as much freedom of movement as possible in order to simulate human motion adequately. The model is a replica of the articulated models usually used in drawing classes (fig. 13). At present, its simple "body" is divided into various main segments, followed by an anatomical simplification of the limbs, so that each section corresponds to a drawing function written in AutoLISP code.

The geometry used in the Humanoid puppet is a set of elementary shapes with a substructure that is represented by drawing: a "circle" for "connectors and head"; and a "rectangle" for "straight limbs", so that the former are responsible for the rotation of the latter, just as the mechanisms of a real human body function.

Each of these functions is a drawing operation defined by three set parameters: insertion point by coordinates (x and y), width of the shoulders (width) and height of the body (height). The parameter percentages have been introduced by approximation to the proportions between limbs, so that they can be manipulated to draw as many variations of "body types" as possible.

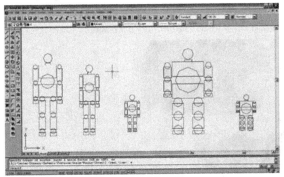

Fig. 12. Output of "Genera(c)tiveHumanoidLifeForm" AutoLISP code: Male, Female, Child, OverAveragePerson and UnderAveragePerson. Image by M. Piedade Ferreira

Fig. 13. Humanoid drawing model used as a reference for digital parametric puppet

By studying the motion of the human body in space, we are trying understand how the human being, as an organism with specific psychophysiological characteristics and an innate capacity for interaction and spatial definition, generates space through movement and gives shape to his habitat "in the same way a bird shapes its nest, with the movements of its body" [Bachelard 2005: 113]. While simulating spatial generation through movement and body language, we hope to find the expression of the architectural genotype and phenotype, the matrix or source of a natural and corporeal architecture based on instinctive human gestures of protecting the body, many of which are universal and the basis of architectural archetypes. To achieve these simulations, at a more advanced stage the program will animate avatars to work as responsive agents in the generation of architectural space, and algorithms will be developed to interpret their movements while performing activities connected with future experiences of the spaces they ought to generate. These algorithms will determine how such space will be generated in terms of shape: in some cases the interpretation will be literal and the form of these spaces a direct transcription of avatar movements whereas in other cases, the interpretation will be more refined and the shape of the spaces generated will only bear a vague relation to these movements.

At this stage, the program that has been developed is a literal interpretation of the avatar's movements and generates animations, or choreographies, by recursion of instructions or transformation rules, in this case, rotation and copy. Shapes and spatial relations are generated by rotating the limbs and the sequential copying of each of the results. The overlapping of the program's output was chosen as an analogy to the Futurist synthesis used to study movement and simultaneity, a technique also developed in film both by the Russian Constructivists such as Popova and Vertov, and the Italian Futurists such as Balla (figs. 14 and 15) and Boccioni (fig. 16). Boccioni's work is a particularly good example of how form might be generated by an interpretation of human movement.

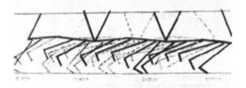

Fig. 14. Giacomo Balla's technical study for *Girl Running on a Balcony*, 1912

Fig. 15. Balla's *Dynamism of a Dog on a Leash*, 1912. Albright-Knox Art Gallery, Buffalo, New York

Fig. 16. Umberto Boccioni's *Forme uniche della continuità nello spazio* (1913, Milano, Civico Museo d'Arte Contemporanea) explores the plastic continuity of the movement of the human body in space

Eadweard Muybridge's work is another important reference in our work. Muybridge was considered by many to be the father of cinema and a pioneer in the study of animal locomotion, using a self developed photographic multiple exposure system called the "Zoopraxiscope" [Zoö + Gr. praxis a doing, an acting (from to do) + scope.] that allowed images to be projected sequentially onto a screen, thus generating animations. Our idea is to create software that facilitates the composition of complex choreographic movements and subsequently the generation of architectural space from such choreographies using algorithms to interpret movement forms and translate them into architectural space. Muybridge's work may be used to depict the functioning of our program. The first step in our software is to animate an avatar to simulate a human moving through space: a person entering home and hanging her coat on a rack, an acrobatic dance performed by someone in a disco listening to his favorite music, or the jump of an Olympic athlete in a stadium. The second step is to use an algorithm to interpret this movement according to some relevant criteria and obtain, for instance, a sequence similar to Muybridge's "Animal Locomotion," Plate 165 (Jumping and pole vaulting) (fig. 17, top). This intermediate output might then be used to generate form using the algorithm to overlap image stills from the sequence (fig. 17, bottom).

The tool we are designing, therefore, will ultimately be used as a dynamic stage set, where parametric avatars or puppets are commanded by the designer and used as flexible templates in the generation of space, producing geometry by their motions, recorded on geometrical surfaces that will be codified in a mathematical language. The quality of the result will depend on the criteria adopted by the designer when choosing characters for the bodies and composing their movements in space. The designer will be working as a director or puppeteer on a customized stage where the characters will perform actions according to his rules or criteria and, by printing the resulting shapes by rapid prototyping, the result will be a "corporeal" materialization of a virtually generated sequence of actions. Rees' work "Putto 8 2.2.2.2" (fig. 18) is useful in illustrating this idea. His work shows the process of using rapid prototyping for making physical form from shapes created by simulating body movement. Rees used software to compose humanoid shapes that were then digitally fabricated, but the process of composing such shapes was not automated. This process was, therefore, time-consuming and the complexity of the composed shapes limited. Our work draws inspiration from his works but it goes one step further by automating the process of generating such shapes. Our idea is to create shape grammar-based software that eases the composition of complex choreographic movements by interpolating between choreographic postures defined as key steps in a movement sequence.

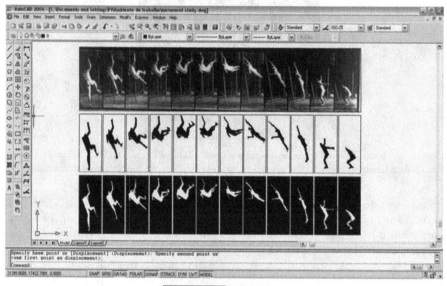

Fig. 17. **Simulation of** the form generation process in the "Genera(c)tiveHumanoidLifeForm" under development. Eadweard Muybridge's "Animal Locomotion," Plate 165 (Jumping and pole vaulting) of 1887 (top) illustrates a possible intermediate output of this program. Image stills in the sequence are then overlapped to obtain two different results (bottom). Images by M. Piedade Ferreira

Fig. 18. Michael Rees's "Putto 8 2.2.2.2, 2003", explores through digital modelling and rapid prototyping the virtual and material composition of (im)possible bodies. Source: http://home.earthlink.net/~dadaloplop/michael_rees.html. Reproduced by permission of the artist

The Grammar of Movement – Rules for a performance machine

> *Since the art of the actor is the art of plastic forms in the space of a stage, he must study the mechanics of his body* [and] *train this material so that it is capable of executing instantaneously those tasks which are dictated externally*
>
> Meyerhold, The Actor and Bio-Mechanics
> [Campbell, Lynton et al. 1971: 80]

In 1909, Marinetti's first Futurist manifesto was published in Russia, providing Russian artists with a powerful weapon against the art forms of the past. Futurist ideals were adapted to support a cultural revolution that started with Mayakovsky's 1912 quasi-futurist manifesto "A Slap in the Face of Public Taste" and generated many artistic movements that would grow in the twenties, such as Suprematism, Rayonism and Constructivism, the latter seen as an ethical proclamation of a social and political revolution that would have a major influence on the development of the aesthetics of the Modern Movement, especially Le Corbusier's architecture. Constructivist artists believed that to surpass academism in the arts, speculative activities such as painting should be abandoned and artists should use real space and real materials in performances of what they called "production art".

This kind of non-conventional theatre combined many kinds of performance arts such as circus arts, puppetry or music hall, and the movement of the human body was studied with actors on stage, using methods and practices such as the eurhythmics of Emile Jacques-Dalcroze, the eukinetics of Rudolf von Laban and Meyerhold's Bio-Mechanics. Meyerhold would become, in 1921, the Director of the State Higher Theatre Workshop in Moscow, where he developed bio-mechanics as a system of rules for actors, based on physical discipline and self-awareness. This allowed for a new dynamic style of theatre and, in the words of Meyerhold, the actor could base his art on scientific principles, transforming the entire creative act into a conscious process which would help him use his body's means of expression correctly to arouse the emotions of the spectator, inducing him to share his performance [Campbell, Lynton et al. 1971: 60-81]. "The Magnanimous Cuckold" (fig. 19), 1922, was Meyerhold's first staged Constructivist performance, in which the actors were carefully placed and conducted by the director in a rhythmic production. The stage set was designed by Popova and consisted of a set of interconnected apparatus operating as a dynamic extension of the actors' movements in space.

Fig. 19. Stage set for Meyerhold's "The Magnanimous Cuckold" designed by Liubov Popva, 1922

Merging the actor with the scenography was also one of the intentions of the Futurists, who wanted to make a synthesis of sound, scene and gesture to create a psychological synchronism in the soul of the spectator, compressing into a few minutes of improvisation innumerable situations, sensibilities, ideas, sensations, facts and symbols (see [Goldberg 2006: 26]). The Futurists also outlined rules for the movement of human bodies on stage, advising the actor to gesticulate geometrically, in a draughtsman-like topological manner, synthetically creating cubes, cones, spirals and ellipses in mid-air. To explore these concepts, many futurist artists constructed and performed with "Übermarionettes" or life-sized puppets whose geometry would allow the idea of a dynamic sensation made eternal through mechanization to be explored.

Although these concepts were developed to create a sense of "gripping" or immersing the spectator in the actor's performance, it is possible to establish an analogy between this kind of experience and an architectural space serving as a dwelling that is generated by a kind of actor, in this case, the tool we are developing, a humanoid puppet in a flexible digital setting. In other words, if the movements and expressions of an actor on stage or film can arouse the emotions of the spectator and if these movements can be translated into a material architectural space, the dweller may sense the actions and emotions that generated the space.

Following Meyerhold's scientific methodology, in order to achieve a performance, or a space, that can "grip" the spectator or the dweller, it is necessary to establish a system of rules that govern the actor's body language as a communication tool. According to Schlemmer's performance theory, these rules must be considered decisive in the transformation of space through the action of the human body: the rule of the circumscribing cubit space; the functional laws of the human body in its reaction in space, the rules of human locomotion in space and the metaphysical forms of expression. Leonardo da Vinci, in his treatise *On Painting*, presents a set of rules that allow for the appropriate depiction of the human body so that the postures of the characters in the paintings represent motion in their bodies and especially in their minds (see [Kemp 1989: 120, 144-146]).

As previously stated, in a further phase our tool will allow the action and reaction of dwellers, as responsive avatars in digitally constructed environments, to be simulated. In this case, the shape of the architectural spaces will be intentionally manipulated by the user to create a certain psychophysiological impact. This can be achieved by changes in light, colour, texture or sound and by placing the architectural elements in ways that will make the user follow certain paths in certain rhythms, similar to the method used by Meyerhold in the setting of "The Magnificent Cuckold". By "composing" the dweller's movements in space, bearing in mind that he is always constructing an individual and subjective experience, we hope to achieve a synthesis of sensory stimulation that will immerse him in architectural space on a subliminal or subconscious level. With this approach, we are trying to test through behaviour simulations whether it is possible to establish a calculated body/mind game with the dweller that works in the same way as film or computer game experiences, and to see if this methodology can be useful in generating designs that can provide the potential for individuation and a more corporeal experience of architectural space.

In this sense, our first operational approach to shape grammars aims for an understanding of how the dynamics of the human body codified by rules can be used as a tool to generate architectural space and thus corporeal experience. We hope that the systematic use of rules will help us understand what is required to reproduce the human body in motion and also how to generate sequences of actions and architectural space by simple or combined motions that have a specific impact on the human body in a holistic sense.

In the grammar of movement that is being developed, one or more shape rules enabling the body to move from one position to another correspond to each figure. The starting position is the static position of a standing human and this brief set of rules can be used to generate choreographies. This language will be extended to encompass other figures, thereby enlarging the universe of actions. Subsequently, it will be possible to define several sublanguages, for instance one for dance, another for yoga, another for common daily tasks and so on, bearing in mind that these languages are subsets of the language of human movement. Our simulation tool will be developed according to the rules or set parameters codified by the rules in the grammar.

In the grammar developed so far, the chosen movements are yoga postures, namely a set of yoga postures due to their anthropological value as some of the most primitive expressions of human consciousness of the cosmos and also their psychophysiological impact on the human body. They have also been chosen for their symbolic elemental geometry and their capacity to generate complex choreographies by allowing a great variety of combinations of elements and coherent results from random or chosen rules. We will now present a simple rule description, and give some examples of generated designs, or choreographies, describing different levels of complexity. Twelve rules have been developed so far, and the grammar is a composed grammar in order to achieve a more extensive description of the transformations which occur. Each rule presents two views on each side, a front view on the left and a side view on the right (fig. 20).

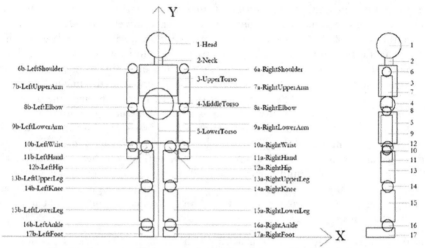

Fig. 20. Grammar of Movement Initial Shape. Image by M. Piedade Ferreira

The initial shape is an output of the Genera(c)tiveHumanoidLifeForm program, a parametric humanoid based on the human body's system of proportions. To facilitate the application of the rules for each geometrical shape that composes the "body" each one was identified by a number, as shown in the diagrams in fig. 21. Each rule moves a set of these "body" parts following a proper procedure. Consider, for instance, Rule R3 in fig. 22. To rotate the right arm centring on the shoulder until it becomes parallel to the ground, one of the movements included in this rule, the procedure is as follows: with centre in 6a, rotate 90° the "body" shapes 6a ; 7a ; 8a ; 9a ; 10a ; 11a. This rule also moves the left arm and the two legs using similar procedures. Another rule, R12, is shown in fig. 23. These rules encode a vocabulary of choreographic postures and can be used to introduce them in a chosen sequence on strategic points in space, thereby defining a specific choreography. Fig. 23 shows an example of a choreography obtained by random application of the rules. The set of all possible choreographies that can be generated from the rules form a language of choreographic designs. Future work will be concerned with two aspects. The first is to impose restrictions on the rule application sequence thereby creating higher level rules that can be used to define sublanguages. The other is to write rules to interpolate between postures and originate movement.

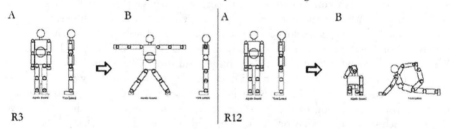

Figs. 21 (left) and 22 (right). Examples of Rules R3 and R12, respectively. Images by M. Piedade Ferreira

Fig. 23. Choreography generated by random application of the rules.
Images by M. Piedade Ferreira

Final considerations

The work described in this paper aims to demonstrate our efforts to gather a consistent body of knowledge on what may be the decisive parameters needed to start to develop a design methodology which, by using a carefully selected system of rules and tools capable of encompassing a large amount of complex parameters, can generate objects that allow for a more intuitive, intelligent and emotional interaction between them and dwellers. Our aim is that this "corporeal" view of architecture will be able to operate in a prophylactic or therapeutic way capable of confronting some of the adversities generated by the impact of today's technology on the human body, by stimulating its sensory system holistically. We have also tried to demonstrate how the incorporation, from an historical and critical standpoint, of other systems and methodologies, technological, artistic or scientific for example, can be very useful in the development of a new tool, since they have already opened up the path required for the work but have stopped, due to a lack of means or simply due to changes in technological, cultural or political conditions. In addition, this research has allowed us to reflect on the importance of a structured and reasonably reliable system of rules or criteria that can support the use of such tools, such as the simulator we are trying to develop, bearing in mind that, as Le Corbusier pointed out in *"Des yeux qui ne voient pas"*, a tuned instrument is not enough to generate harmony, it is necessary to use standards, "… the product of logic, of analysis and painstaking study; they are evolved on the basis of a problem well stated. In the final analysis, however, a standard is established by experimentation" [Le Corbusier 1954: 33].

Notes

1. Cabral De Mello's thesis *A Arquitectura Dita / Anamorfose & Projecto* [2007] adapts to architecture the equivalent concept coined by Saumjan [1965].

References

ALEXANDER. Christopher. 1965. "A city is not a tree", In *Architectural Forum*, Vol 122, No 1, April 1965, (Part I), Vol 122, No 2, May 1965, pp. 58-62;

BACHELARD, Gaston. 2005. *A Poética do espaço* (1957). António de Pádua Danesi. São Paulo. Martins Fontes.

BERGSON, Henri. 1991. *Matter and Memory*. (1896). New York. Zone Books.

CABRAL DE MELLO, Duarte. 2007. *A Arquitectura Dita / Anamorfose & Projecto* (PhD thesis). Technical University of Lisbon.

CAMPBELL, Robin, Norbert LYNTON, et. al. 1971. *Art in Revolution: Soviet Art and Design since 1917.* Catalogue from the exhibition in the New York Cultural Center. 9 September to 31 October 1971. New York. London. Arts Council.

GOLDBERG, RoseLee. 2006. *Performance Art, from Futurism to the Present* (1988). "word of art". London. Thames & Hudson.

HILLIER, Bill; HANSON, Julienne. 1996. *The Social Logic of Space*. Cambridge. New York. Cambridge University Press.

KEMP, Martin. 1989. *Leonardo On Painting*. Martin Kemp, Margareth Walker, selec. and trans.. Newhaven and London.Yale University Press.

GOEL, Nitin. 2001. *Shiva as Nataraja – Dance and Destruction In Indian Art.* Ed. Nitin KUMAR. ExoticIndianArt Pvt Ltd. http://www.exoticindiaart.com/article/nataraja. Last accessed 16 November 2010.

LE CORBUSIER. 2000. *The Modulor.* 1954. Peter de Francia and Anna Bostock, trans. Rpt. Basel: Birkhäuser.

MCEWEN, Indra Kagis. 2003. *Vitruvius – Writing the Body of Architecture.* Cambridge, Massachusetts, London, England. The MIT Press.

MERLEAU PONTY, Maurice. 2002. *Phenomenology of Perception.* (1945). New York, London: Routledge.

PALLASMAA, Juhani. 2005. *The Eyes of the Skin: Architecture and the Senses.* West Sussex England. Wiley Academy Press. Wiley & Sons.

QUEYSANNE, Bruno. 1987. Penser l'Architecture c'est Penser Autrement. Pp. 95-98 in *Mesure pour Mesure, Architecture et Philosophie*, n. sp. Cahier du CCI, Centre de Création Industriel. Paris: Centre George Pompidou.

RASMUSSEN, Steen Eiler. 1964. *Experiencing Architecture.* (1959). Cambridge, Massachusetts, London, England. The MIT Press.

SAUMJAN, Sebastian Kontantinovic. 1970. Cibernética e Língua. Pp. 129-144 in *Novas Perspectivas Linguísticas*, M. Lemele and Y. Leite, orgs. Petrópolis, Rio de Janeiro: Editora Vozes.

SCHLEMMER, Oskar 1987. *Escritos sobre Arte: Pintura, Teatro, Danza, Cartas y Diarios.* (1977) Ramón Ribalta, trans. Barcelona: Paidos Estetica.

SILVÉRIO MARQUES, Manuel Barroso. 1990. *Modularity, Mind and Brain Theory – an essay on Fodor's Theory of the Mind* (1985). In Controvérsias Científicas e Filosóficas. Ed. GIL, Fernando. Lisbon. Editorial Fragmentos, Lda. pp. 159-187.

PREZIOSI, Donald. 1979. Semiotics of Built Environment: Introduction to Architectonic Analysis. Bloomington: Indiana University Press.

VARELA, Francisco J.; Thompson, Evan; Rosch, Eleanor, co-authors. 1991. *The embodied mind: cognitive science and human experience.* Cambridge, Massachusetts, London, England. The MIT Press.

ZEIZEL, John. 2006. Inquiry by Design - Environmental / Behaviour / Neuroscience in Architecture, Interiors, Landscape, and Planning. (1981). Revised Edition (New York, London, W.W. Norton & Company).

About the authors

Maria da Piedade Ferreira is an architect. Her transdisciplinary Ph.D. research is focused on the holistic study of the human body, combining architecture with performance arts, new media, cognitive sciences, human-machine interfaces and anthropology.

Duarte Cabral de Mello (Ph.D.) is an architect and Assistant Professor at the Faculty of Archicture of the Technical University of Lisbon. His work has been selected for exhibitions of Portuguese Architecture in Lisbon (1986, 1987, 1989, 1998), Porto (1991), Brussels (1991) Tokyo (1992), New York (1994) and Munich (1997), and has been published in professional journals including *Arquitectura, Architecti, Arquitectos, Domus, Architecture d'Aujourd'hui, A&V.* His main ongoing research area is the creativity and ethics of architectural and urban design.

José P. Duarte holds a B.Arch. (1987) in architecture from the Technical University of Lisbon and an S.M.Arch.S. (1993) and a Ph.D. (2001) in Design and Computation from MIT. He is currently Visiting Scientist at MIT, Associate Professor at the Technical University of Lisbon Faculty of Architecture, and a researcher at the Instituto Superior Técnico, where he founded the ISTAR Labs - IST Architecture Research Laboratories. He is the co-author of *Collaborative Design and Learning* (with J. Bento, M. Heitor and W. J. Mitchell, Praeger 2004), and *Personalizar a Habitação em Série: Uma Gramática Discursiva para as Casas da Malagueira* (Fundação Calouste Gulbenkian, 2007). He was awarded the Santander/TU Lisbon Prize for Outstanding Research in Architecture by the Technical University of Lisbon in 2008. His main research interests are mass customization with a special focus on housing, and the application of new technologies to architecture and urban design in general.

Alexandra Paio

Departamento de
Arquitectura e Urbanismo
ISCTE-IUL
Lisbon University Institute
Av. das Forças Armadas
1649-026 Lisbon PORTUGAL
alexandra.paio@iscte.pt

Benamy Turkienicz

SimmLab - Laboratory for the
Simulation and Modelling in
Architecture and Urbanism
Universidade Federal do Rio
Grande do Sul
Rua Sarmento Leite 320
Porto Alegre, RS, BRAZIL
CEP 90020-150
benamy.turkienicz@gmail.com

Keywords: shape grammar;
Portuguese urban design;
proportion; symmetry

Research

An Urban Grammar Study: A Geometric Method for Generating Planimetric Proportional and Symmetrical Systems

Presented as a poster at Nexus 2010: Relationships Between Architecture and Mathematics, Porto, 13-15 June 2010.

Abstract. This paper is part of ongoing research to explore (a) descriptive and generative potentials of shape grammars to unveil the order of urban morphological complexity underlying Portuguese treatises, and (b) the knowledge embedded in Portuguese urban cartographic representations produced from the sixteenth to the eighteenth centuries for purposes of architecture and military engineering. The shape grammar made it possible to infer the grammatical rules of composition based only on the written descriptions provided by the Portuguese treatises, as well as to decode the geometry of ideal urban plans and built cities. This paper suggests that these historical cities, both built and ideal were based on a structured knowledge-based process from where it is possible not only to retrieve a generative parametric urban grammar but also to construct a computational model, UrbanGENE, capable of iteratively generating Portuguese planimetric proportional and symmetrical urban systems. The paper suggests that the knowledge and computational tools achieved could be successfully deployed in the teaching and learning of architectural history.

1 Introduction

In the last decade, urban history has been very important for the dissemination of knowledge about Portuguese urban design. These studies have developed hypotheses of morphological continuity in Portuguese urban design from the sixteenth to the eighteenth centuries based on shared composition rules and procedures. Architects and historians have demonstrated that Portuguese urban design evolved from planned, regular and erudite principles of precision and standardization [Araujo 1992, 2000; Horta-Correia 1997; Carita 1999; Teixeira & Valla 1999]. This is in contrast to previous studies in which where the lack of planning rules was considered a pervasive characteristic of Portuguese urban design [Holanda 1936; Azevedo 1956; Smith 1955; Santos 1968].

Along with these two contradictory assumptions, there are other issues to be tackled. One of these is related to the description of the process mediating the theoretical knowledge held by urban makers and, at the same time, the representation of this knowledge in urban historical cartography produced from the sixteenth to eighteenth centuries.

To describe this process, we have to infer the genetic and generative principles of Portuguese urban design, as well as understand the form-making logic. The understanding of relationships between morphological elements, categories and classes of

---

DOI 10.1007/s00004-011-0064-6; *published online* 9 March 2011

basic relations, supports the recognition of the city as a structured order with repeated composition rules.

The historical documentation suggests that Pythagorean-Euclidian geometry is a crucial ingredient for understanding Portuguese urban design-thinking and urban design-making. To unveil this genesis, a descriptive method, shape grammar [Stiny & Gips 1972; Stiny, 1980] has been adopted. As a method, shape grammar supports the analysis of the form-making logic and has proved to be powerful in the shape analysis, description, interpretation, classification, evaluation and generation of design languages [Stiny & Mitchell 1978, 1980; Knight 1981, 1989; Buelinckx 1993; Brown 1997; Turkienicz et al, 2005, 2006, 2007; Colakoglu 2005] and to provide a gateway to define parametric urban shape grammars [Beirão & Duarte 2005; Steino 2005; Duarte et al, 2006].

The present shape grammar study permitted the analytical/descriptive morphological decomposition of the sixteenth to the eighteenth century's historical cartography of Portuguese urban designs and has helped to unveil the geometric principles that characterize Portuguese treatises and practice in this period. As it is known, shape rules [Stiny 1980] have the potential to bridge the gap between traditional drawing techniques and modern computational methods of urban design. Following this potential, it is possible to devise a geometric method for the automatic generation of planimetric proportional and symmetrical urban systems. Taking this idea as departure point, the goal of this research is to develop a parametric urban grammar computational model to generate new Portuguese towns during the sixteenth to eighteenth centuries. The model should enable the grammar's user to interpret, manipulate and generate simultaneously various geometric constructions as representations of Portuguese historical urban designs.

This paper has four sections. The first section shows, briefly, some traditional views based on iconographic evidence and historical data. The second section describes the process of inferring and structuring the urban grammar. The third, introduces a geometric method for generating Portuguese planimetric proportional and symmetrical urban systems. The final section discusses the partial results of the research.

2 Knowledge-based Portuguese urban design

The complexity of Portuguese urban design from the sixteenth to eighteenth centuries has been the subject of different authors [Ribeiro 1962; Horta-Correia 1997; Lamas 1993], under different analytical methods. Bodies of knowledge as diverse as economics, sociology, history, geography and architecture have been used to interpret Portuguese urban form of the cited period from different point of views.

Three basic concepts have emerged from the explanation and description of the historical Portuguese planned city. The first concept arises from the ideal city models present in Portuguese treatises. Several scholars have argued, interpreting the knowledge present in available treatises, that the existing theory and the pragmatism from Portuguese urban makers constituted predominant factors behind the planning and the construction of the Portuguese city [Chicó 1956; Carita 1999; Bueno 2003; Horta-Correia 1997; Moreira 1982; Araujo 1992, 1998; Murteira 1999; Paio 2007, 2009].

The second concept identifies morphological elements and its correspondent structural relationships [Lamas 1993; Rossa 2002; Teixeira & Valla 1999; Reis Filho 1962; Rhoden1999]. For these authors, the identification of discrete morphological elements (street, square, institutional buildings, blocks, lots) constitutes the basis for the analysis, description and the understanding of categories of basic relations and, additionally, for the understanding of the city as the result of a structured order with repeated rules.

The third and final concept begins with the influence of the topographic features [Ribeiro 1962; Gaspar 1969; Salgueiro 1992; Amaral 1987; Teixeira & Valla 1999] in the genesis of Portuguese cities. The city is then described according to the ability of the Portuguese urban makers to adjust or to superimpose a planed urban geometric structure onto the chosen site to build the city.

These three approaches advocate that Portuguese urban morphological order combines theory and the urban maker's knowledge: knowledge not only to design, but to actually build cities. Authors argue that it is precisely this know-how that allowed Portuguese urban makers to successfully manipulate geometric structures and morphological relationships to promote the city as a structured system of forms and spaces. The awareness of geometry has given birth to pragmatic procedures, enabling the Portuguese effectively respond effectively to a wide variety of conceptual problems and to use urban design to efficiently control their territory according to a set of pre-defined objectives.

Authors were also concerned about linking particular solutions to individuals or excessively fragmenting or classifying urban designs according to historical periods [Teixeira 2008]. Portuguese urban design studies [Rossa et al. 2006] have tried to structure the whole picture without pinpointing general compositional rules and procedures. As Duarte put it "the traditional studies in urban and architecture history only describe a style without saying how it is possible to generate new projects in that style" [2007:20]. In other words, Portuguese urban history lacks of a description based on genetic and generative principles explaining a plurality of design solutions that characterized the sixteenth to eighteenth centuries. This paper represents an effort to overcome this problem by using the descriptive and generative morphological decompositions of shape grammar in order to unveil genotypic similarities beyond the apparent diversity present in the *Corpus* (fig. 1, fig. 2).

3 The Portuguese urban morphogenetic genesis

> ... the principles of practical geometry are necessary for Military Architecture intelligence and *Fabrica* ... who was not good as a geometer and as an arithmetic, will fail without knowing his mistakes and only will discover them when the object is built ...
>
> Pimentel, 1680 [1993: 3].

In order to analyze the design knowledge embedded in Portuguese cities, we defined a corpus of thirty Portuguese treatises (practical geometry, military architecture, military engineering, etc.), used in military classes of Portuguese schools from the sixteenth to eighteenth centuries (fig. 1), and a corpus of seventy-five samples of Portuguese urban historical cartography (fig. 2).

The analysis of the treatises has furnished, in two steps, the primary genetic and generative principles of Portuguese urban design. The first step disclosed a Pythagorean-Euclidean geometric genesis, the geometric operations necessary to build the syntax of the Portuguese urban plans. The second step has revealed the Vitruvian urban genesis present in the treatises, the theoretical urban models.

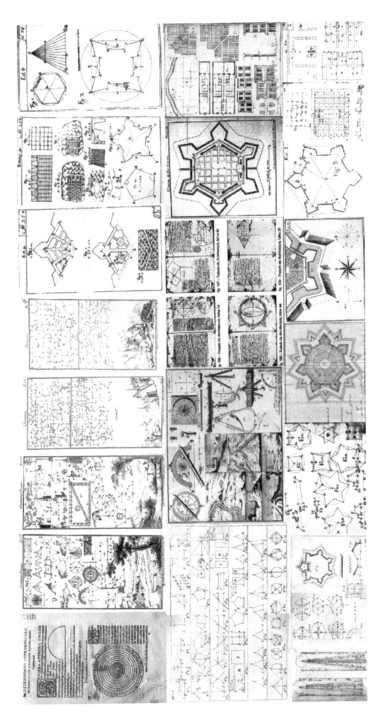

Fig. 1. The corpus of thirty Portuguese treatises (practical geometry, military architecture, military engineering, etc), used in military classes of the Portuguese's schools from the sixteenth to the eighteenth centuries

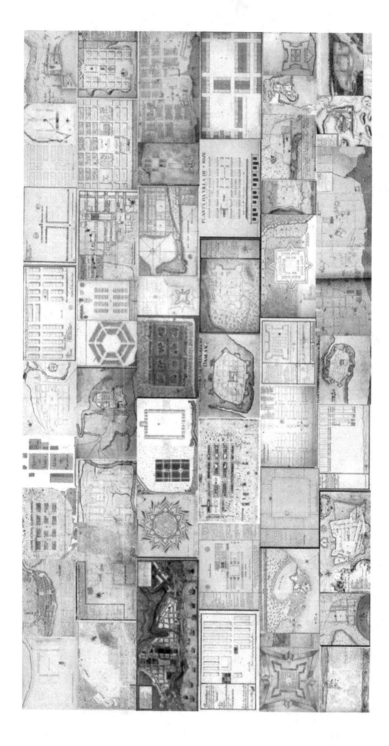

Fig. 2. The corpus of seventy-five samples of Portuguese urban historical cartography from the sixteenth to the eighteenth centuries

The knowledge embedded in the treatises confirmed the ability of Portuguese urban planners to configure geometric proportional structures along with invariant features of many Portuguese planimetric urban systems. These invariances have made it possible to develop a parametric shape grammar capable of capturing the genetics of Portuguese urban plans from the sixteenth to the eighteenth centuries.

3.1. Pythagorean-Euclidean geometric genesis. The analysis of Portuguese treatises and compendiums helps us to identify the structure of the rationale of urban planning taught to apprentices and, additionally, to decode the urban design logic supported by this rationale. In the treatises present a particular way of conceptualizing and systematizing knowledge.

Geometry, in its operative dimension, becomes a preparatory science, a mental discipline, an abstract discourse enabling models to be built from visual reasoning processes. In this sense, form theory is mathematical in nature and comprises a rich vocabulary, which unambiguously describes geometric forms (such as points, lines, circles, triangles, squares, pentagons, etc.) and geometric qualities (such as symmetry, proportion, hierarchy, rhythm, etc.). As Alberti said "design a firm and graceful preordering of the lines and angles, conceived in the mind and contrived by an ingenious artist" [Alberti 1955: 1-2].

Formal axiomatic Euclidean geometry, that is, geometric constructions based on the compass and an unmarked straightedge, was fundamental for Portuguese urban planners. These were essentials to the abstract mental process and visual reasoning in the construction of formal models, exact representation and construction on the site.

3.2. Vitruvian genesis. Portuguese treatises reveal a strong component of classical, Vitruvian origins [Carvalho 2000; Bueno 2003; Paio 2007]. Pimentel's remarkable Portuguese treatise on fortification design and city planning, *Método Lusitânico de Desenhar as fortificações das Praças Regulares e Irregulares* (1680), shows the complex relationship between abstract, formal considerations and pragmatic military issues. This military compendium reveals not only a confident assimilation of Vitruvian planning ideas but also many other European military treatises from the seventh century [Carvalho 2000] (fig. 3). Pimentel described in detail the important components, relations, and dimensions of the ideal city, but didn't provide any illustration of it. To "formalize" the written description, a formal language – shape grammar – has been used, making possible a graphic interpretation of the text, which can then be used to generate of some urban plans (fig. 4, fig. 5).

Based on Pythagorean-Euclidian structure it was possible to deduce a geometric method of design starting from a center (the point of the compass), a direction (essential to define symmetry), and a limit (a circle), which expresses domain and periphery [Critchlow 2001] (fig. 4). Alberti commented, "... Platforms should all terminate within a circle, and indeed from a circle is the best way of deducing them" [Alberti 1955: 194]. A circle that is divided in three equal parts defines the generative shape of the planimetric system. After using a geometric transformation operation, rotation, it was possible to establish the beginning of two triangle-based proportions, one based on the hexagon and the other based on the root-three rectangle. These two triangle-based proportional structures would ensure a repetition of ratios and shapes throughout the design [Kappraff 2002; Scholfield 1958], as well as adapt to different scales and hierarchies. These two options will generate two basic planimetric proportional and symmetrical systems, a grid whereby urban elements such as streets, squares, urban blocks, and main buildings emerged (fig. 4, fig. 5).

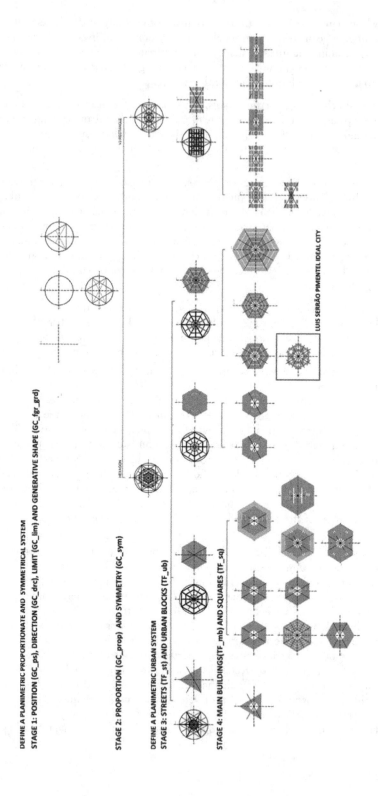

Fig. 3. Tree diagram to illustrate the theoretical urban models present in European military treatises from the seventh century

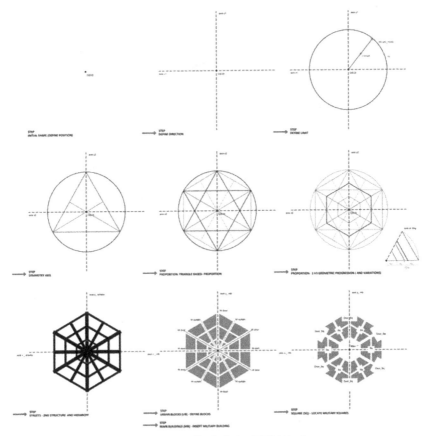

Fig. 4. Pimentel's ideal city (TPf_i10_1)

3.3. Geometric and generative principles. Through the analytical decoding of the corpus' grammar, it was possible to progressively depict geometric and topological attributes and to establish two sets of categories and classes [Mitchell 1998]. These were geometric-configurational (GC) and topological-functional (TF) (Table 1). Each category was divided into four classes. The geometric-configurational (GC) category is strongly Euclidean knowledge-based and is constituted by seven different elements: Position (GC_ps), Direction (GC_drc), Limit (GC_lim), Generative Shape (GC_gene.shape), Diagonal (GC_dng), Proportion (GC_prop) and Symmetry (GC_sym) [Paio and Turkienicz 2009]. The topological-functional (TF) category is related to urban elements: Streets (TF_s); Urban Blocks (TF_ub); Main Buildings (TF_mb) and Squares (TF_sq) (Table 2). The two categories are related so as to make it possible to associate the category's four classes to each other. Classes of the geometric-configurational (GC) category were used to generate symmetrical and proportional layouts where the geometric structure has been deployed to generate streets and urban blocks. The definition of the point (geometric center) and the direction (vertical and horizontal axes) permitted operations such as rotation and, further on, the positioning of the church (ch), town hall (th), or military buildings (mlt) and the location of one or more squares (sq).

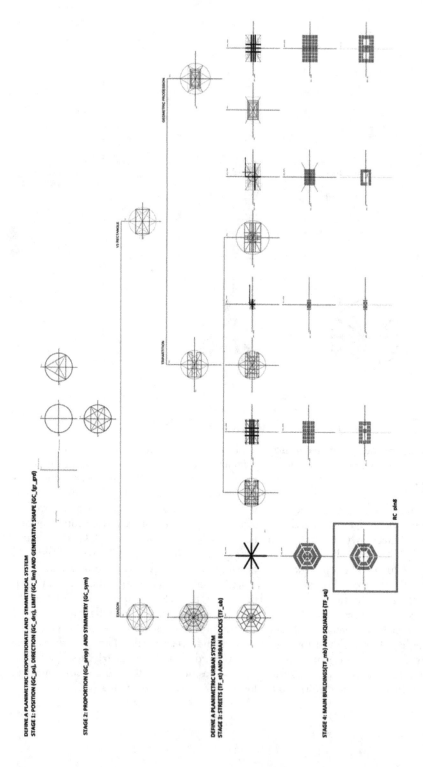

Fig. 5. Tree diagram to illustrate the generation of some Portuguese urban plans from the Pimentel's ideal city

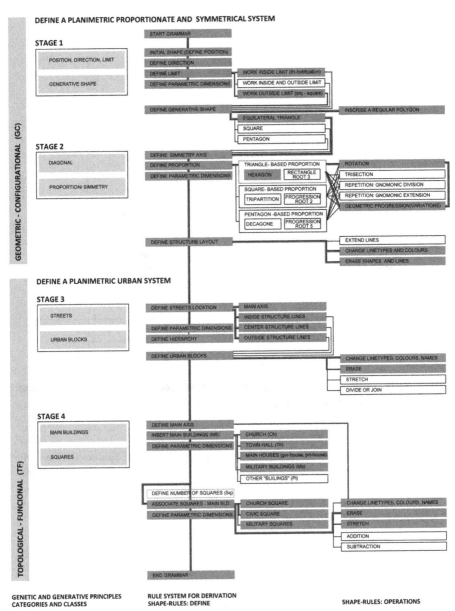

DEFINE A PLANIMETRIC PROPORTIONATE AND SYMMETRICAL SYSTEM

STAGE 1
- POSITION, DIRECTION, LIMIT
- GENERATIVE SHAPE

START GRAMMAR
INITIAL SHAPE (DEFINE POSITION)
DEFINE DIRECTION
DEFINE LIMIT — WORK INSIDE LIMIT (frt-fortification)
DEFINE PARAMETRIC DIMENSIONS — WORK INSIDE AND OUTSIDE LIMIT
WORK OUTSIDE LIMIT (srq - square)

DEFINE GENERATIVE SHAPE — INSCRIBE A REGULAR POLYGON
EQUILATERAL TRIANGLE
SQUARE
PENTAGON

STAGE 2
- DIAGONAL
- PROPORTION/ SIMMETRY

DEFINE SIMMETRY AXIS
DEFINE PROPORTION
DEFINE PARAMETRIC DIMENSIONS

TRIANGLE- BASED PROPORTION — ROTATION
HEXAGON | RECTANGLE ROOT 3 — TRISECTION
REPETITION: GNOMONIC DIVISION
SQUARE- BASED PROPORTION — REPETITION: GNOMONIC EXTENSION
TRIPARTITION | PROGRESSION ROOT 2 — GEOMETRIC PROGRESSION(VARIATIONS)
PENTAGON -BASED PROPORTION
DECAGONE | PROGRESSION ROOT 5

DEFINE STRUCTURE LAYOUT — EXTEND LINES
CHANGE LINETYPES AND COLOURS
ERASE SHAPES AND LINES

DEFINE A PLANIMETRIC URBAN SYSTEM

STAGE 3
- STREETS
- URBAN BLOCKS

DEFINE STREETS LOCATION — MAIN AXIS
INSIDE STRUCTURE LINES
DEFINE PARAMETRIC DIMENSIONS — CENTER STRUCTURE LINES
DEFINE HIERARCHY — OUTSIDE STRUCTURE LINES

DEFINE URBAN BLOCKS — CHANGE LINETYPES, COLOURS, NAMES
ERASE
STRETCH
DIVIDE OR JOIN

STAGE 4
- MAIN BUILDINGS
- SQUARES

DEFINE MAIN AXIS
INSERT MAIN BUILDINGS (MB) — CHURCH (Ch)
DEFINE PARAMETRIC DIMENSIONS — TOWN HALL (Th)
MAIN HOUSES (gvr-house, prt-house)
MILITARY BUILDINGS (Mb)
OTHER "BUILINGS" (Pl)

DEFINE NUMBER OF SQUARES (Sq)
ASSOCIATE SQUARES - MAIN BLD. — CHURCH SQUARE — CHANGE LINETYPES, COLOURS, NAMES
DEFINE PARAMETRIC DIMENSIONS — CIVIC SQUARE — ERASE
MILITARY SQUARES — STRETCH
ADDITION
SUBTRACTION

END GRAMMAR

GEOMETRIC - CONFIGURATIONAL (GC)
TOPOLOGICAL - FUNCIONAL (TF)

GENETIC AND GENERATIVE PRINCIPLES
CATEGORIES AND CLASSES

RULE SYSTEM FOR DERIVATION
SHAPE-RULES: DEFINE

SHAPE-RULES: OPERATIONS

Table 2. Geometric-configurational (GC) and topological-functional (TF) categories

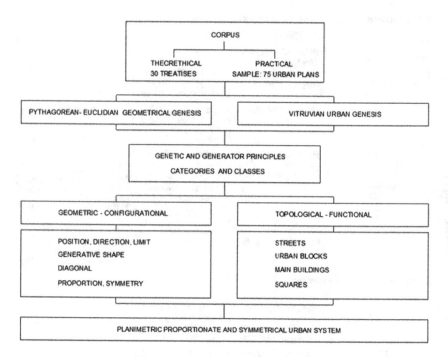

Table 1. Theoretical framework

4 An urban grammar. A geometric method for generating Portuguese planimetric proportionate and symmetrical systems.

The proposed historical urban design grammar is a parametric shape grammar defined in the algebras U12 V12 W12. The generation of an urban derivation develops over four stages: two to generate the planimetric proportional and symmetrical system and two to generate a planimetric urban system (Table 2). These stages are: (1) define position, direction, limit and generative shape; (2) define the rules of proportion and symmetry; (3) define streets and urban blocks; (4) insert main buildings and squares. Each stage has a specific set of shape-rule schemata. These stages are sequential, using a step-by-step process, enabling the generation of plans similar to the Portuguese urban plans from the sixteenth to the eighteenth centuries. Transitions between sequential rules application and stages are controlled by the Shape-Grammar Meta-Language (SGMT) descriptors [Liew 2004]. SGMTs established an alternative method to write grammars for design, introducing seven descriptors for shape grammar language. These explicitly determine the sequence in which a set of rules is applied. At the same time, the method restricts the rule's application through a filtering process using the context to guide the rule matching process [Liew 2004]. The descriptors modify the conditions (rule selection, drawing state, matching conditions and application method) surrounding the process of applying a rule in shape grammar.

4.1. Stages of the urban grammar developing process. The urban grammar presented is still limited in that it only partially describes the range of elements of Portuguese urban plans. Specifically, it represents urban planimetric proportional and symmetrical systems, orthogonal streets (s), urban blocks (ub), main buildings (mb) (churches, town halls, priest's house, governor's house, director's house and military buildings) and squares (sq).

Other elements, such as lots, houses and topographical features, have been omitted. Since the grammar is designed as a sequence of stages, the omitted elements can be inserted into the grammar at a later stage. In order to demonstrate the urban grammar developed so far, an example of the process of deriving a Portuguese urban plan is illustrated and described (fig. 10).

4.1.1. Initial Shape. The initial shape for the Portuguese urban grammar consists of a single labeled point O, with a pair of coordinates (0,0) (rule: GC_0_position) (fig. 6, fig. 10).

4.1.2. Stage 1: Define position, direction, limit and generative shape. In the first stage of the urban grammar, the user defines the basic geometric generative principles of the planimetric proportional and symmetrical system. This stage is composed of seven rules, that permit the user to manipulate dimensional parameters of the Limit and the Generative Shape. The first step inserts the geometric axis1 (x, y) to symmetry and proportion and indicates that the user is in stage 1 (rule: GC_1_direction). The application of rule GC_2_lim_wrkout inserts a circle, which will be divided into equal parts defining the generative shape. In this step the user has parameters to define the circle radius (r_limwrk) and limits to work (lim_wrk): work inside (wrksnd), work out and inside (wrkoutsnd) and work outside (wrkout). These parameters are associated with various labels: Fortification (frt) or Public Square (sq). In this case the parameter used was labeled as sq, which will authorize the user only to work out (lim_wrkout) of the generative shape (gene.shape). The final step inserts the generative shape, in this case an equilateral triangle labeled (sq) (rule: GC_3_gene.shape_trg) (fig. 6, fig. 10).

4.1.3. Stage 2: Define proportion and symmetry. In the second stage of the urban grammar, the user manipulates shape rules to create a symmetric and proportional structure, based on a generative shape selected in stage 1. The sequential steps emulate the similitude to the operation with the compass and the straightedge. In order to clarify the following steps, an example shows set operations (steps) that can be manipulated by the user: Step 1. The application of the rule chg_axis_stage2 carries the derivation into stage 2; Step 2. Define the entire axis (sym_axis_trg1; sym_axis_trg2) of the equilateral triangle symmetry (rule: GC_4_symmetry_axis_trg); Step 3. The user has to decide the type of symmetry: Bi-axial or Multi-axial. In this case user applies the Multi-axial that generates a radial configuration. Step 4. Decide the triangle-based proportions needed to work: hexagon or root-three rectangle. The derivation shows that the user repeated the hexagon based on a variation of a geometric progression 1:√3 (tPG:3). The user has several labeled and parameterized rules to choose from. The user has chosen the repetition-division V (3) dhxg2a = 0.10hxg2 and dhxg3a2 = 0.35 hxg3; Step 5. The regulating structure is defined by lines originating in the triangle-based proportion and results in a particular grid whereby streets (s), blocks (ub) and main buildings (mb) will emerge in the following stage. The result is a consistent geometric structure that regulates the planning of the Portuguese urban layouts at many scales (fig. 7, fig. 10).

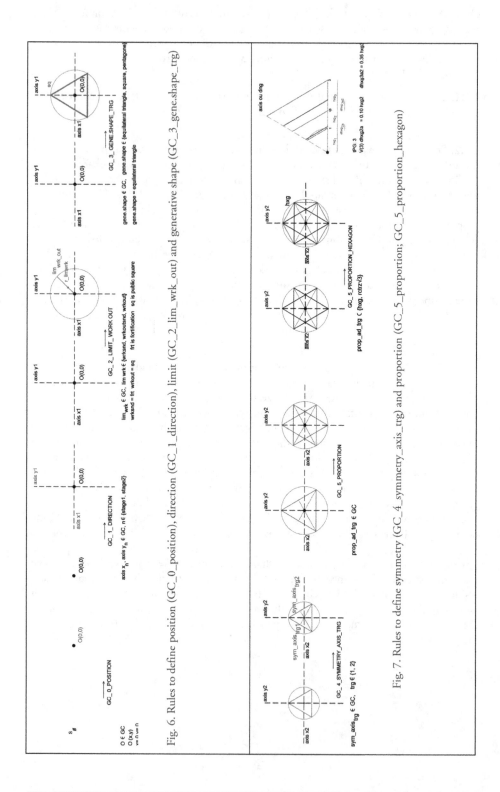

Fig. 6. Rules to define position (GC_0_position), direction (GC_1_direction), limit (GC_2_lim_wrk_out) and generative shape (GC_3_gene.shape_trg)

Fig. 7. Rules to define symmetry (GC_4_symmetry_axis_trg) and proportion (GC_5_proportion; GC_5_proportion_hexagon)

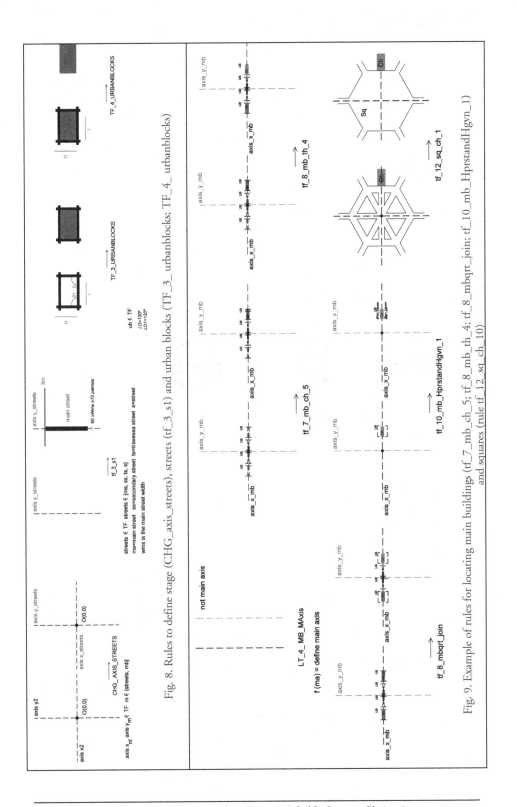

Fig. 8. Rules to define stage (CHG_axis_streets), streets (tf_3_s1) and urban blocks (TF_3_ urbanblocks; TF_4_ urbanblocks)

Fig. 9. Example of rules for locating main buildings (tf_7_mb_ch_5; tf_8_mb_th_4; tf_8_mbqrt_join; tf_10_mb_HprstandHgvn_1) and squares (rule tf_12_sq_ch_10)

DEFINE A PLANIMETRIC PROPORTIONATE AND SYMMETRICAL SYSTEM

STAGE 1: POSITION (GC_ps), DIRECTION (GC_drc), LIMIT (GC_lim) AND GENERATIVE SHAPE (GC_fgr_grd)

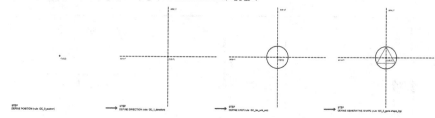

STAGE 2: PROPORTION (GC_prop) AND SYMMETRY (GC_sym)

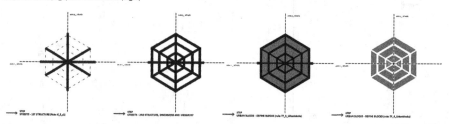

DEFINE A PLANIMETRIC URBAN SYSTEM
STAGE 3: STREETS (TF_st) AND URBAN BLOCKS (TF_ub)

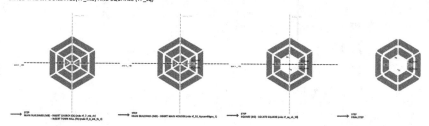

STAGE 4: MAIN BUILDINGS(TF_mb) AND SQUARES (TF_sq)

Fig. 10. The derivation of a Portuguese urban design Sample (RC_pln8)

4.1.4.Stage 3: Insert streets and urban blocks. In the third stage, urban generative elements emerge according to set-rules as follows: Step 1. The application of rule chg_axis_streets leads the derivation into stage 3; Step 2. The application of the rule tf_3_s1 defines first structure streets (s); Step 3. The same rule application defines the final structure of streets. After the user has dimensional parameters to apply (60>Wms>70palmos) and hierarchy (if applicable); Step 4. The recursive application of rule TF_3_ urbanblocks define urban blocks (ub); Step 5. Erase the streets structure and geometric structure no longer necessary (rule: TF_4_ urbanblocks). This final step will highlight the urban blocks grid and the main axis, a fundamental step towards the final stage of the grammar (fig. 8, fig. 10).

4.1.5. Stage 4: Locate main buildings and squares. The application of rule chg_axis_mb leads the derivation into stage 4. The user has, again, set-rules options to locate the main buildings (mb): churches (ch), town hall's (th), governor's house (gvr_house), priest's house (prt_house) and military buildings (mlt) and set-rules to locate the squares (sq) with various formal shapes. All set of shape rules transmit the relation between the main axis and the main buildings (mb) and these with the squares. In this final stage the user has to define main axis to insert the main buildings (mb). The application of rule LT_4_mb_maxis defines the axisx_street as main axis. The derivation shows one way to insert the church (ch) and town hall (th) (rules tf_7_mb_ch_5; tf_8_mb_th_4; tf_8_mbqrt_join). The application of the rule tf_10_mb_HprstandHgvn_1 inserts the governor's house and priest's house. After the rule tf_12_sq_ch_10 locate the associated church square (sq). This stage has set of OPRTV_rules to join, to divide or to stretch that permit manipulate de urban blocks (ub). The final step of the stage and the grammar is the erasure of the main axis (fig. 9, fig. 10).

5 Discussion

This paper suggested a possible application of shape grammar techniques to urban design history. A structural and structured knowledge-based research was essential to recovering the elements of genetic foundations and represent these as constituents of visual reasoning processes and, additionally, to develop a generative parametric Portuguese urban grammar interpreter, UrbanGENE. The results showed that shape grammars can constitute a valuable basis for the understanding of the process of Portuguese urban design. The resulting planimetric system ends up corresponding to a basic compositional procedure supporting the implementation of designs and working as a scale. Once the planimetric system is established, the urban plan incorporates proportions and symmetry in a surprisingly balanced way.

From the preliminary results obtained, it can be said that it is possible to create a useful tool to be used in the learning process of historical urban design. The descriptive and generative character of this tool will allow the user to both interpret and simulate new designs simultaneously based on theoretical knowledge, as well as to manipulate and generate various Portuguese colonial parametric urban design solutions. Since the major part of pre-design stages in urban design are devoted to the study of precedents as strategy for producing new designs, the experience may offer a new incentive to improve the teaching of geometry and urban design precedents in the schools of architecture and urbanism.

References

ALBERTI L. B. 1955. *Ten Books on Architecture*. J. Leoni, trans., J. Rykvert, ed. London: Alec Tiranti

AMARAL, I. 1987. Cidades Coloniais Portuguesas. Notas preliminares para uma geografia histórica. *Povos e Culturas* **2**: 193-214.

ARAUJO, R. 1992. *As Cidades da Amazónia no século XVIII. Belém, Macapá e Mazagão*. Porto: FAUP.

———. 2000. *A Urbanização do Mato Grosso no século XVIII. Discurso e Método*. Ph.D. dissertation, Lisboa: UNL.FCSH.

AZEVEDO, A. 1956. Vilas e cidades do Brasil Colonial: ensaio de geografia urbana. *Boletim da Faculdade de Filosofia e Ciências e Letras*, n. 208.

BEIRÃO, J and J. DUARTE. 2005. Urban Grammars: Towards Flexible Urban Design. *Proceedings of the eCAADe Conference*, Lisbon: 491 - 500.

———. 2007. Urban design with patterns and shape rules. *Proceedings of the 2nd International Seminar on New Town Simulation*: 1-11.

BROWN, K. 1997. Grammatical Design. IEEE Expert, Special issue on *Artificial Intelligence in Design*: 27-33.

BUELINCKX, H. 1993. Wren's language of City church designs: a formal generative classification. *Environment and Planning B: Planning and Design* **20**: 645-676.

BUENO, B. 2003. *Desenho e desígnio: O Brasil dos engenheiros militares (1500-1822)*. Ph.D. dissertation, São Paulo: Faculdade de Arquitectura e Urbanismo da Universidade de São Paulo.

CARITA, H. 1999. *Lisboa Manuelina e a formação de modelos urbanísticos da época moderna (1495-1521)*. Lisbon: Livros Horizonte.

CARVALHO, J. 2000. Luís Serrão Pimentel, O Método Lusitano e a fortificação. MSc. Dissertation, Lisbon: Universidade Lusíada.

CHICÓ, M. 1956. A "cidade ideal" do Renascimento e as cidades portuguesas da Índia. *Garcia de Orta*: 319-328.

CRICHLOW, K. 2001. *Islamic Patterns. An Analytical and Cosmological Approach.* London: Thames & Hudson.

COLAKOGLU, B. 2005. Design by grammar: an interpretation and generation of vernacular hayat houses in contemporary context. *Environment and Planning B: Planning and Design* **32**: 141-149.

DELSON, R. 1979. *New towns for Colonial Brazil. Spatial and Social Planning of the Eighteenth century.* Dellplain Monographs in Latin American Studies 2. Syracuse, NY: Syracuse University.

DUARTE, J.P., G. DUCLA-SOARES, L. CALDAS and J. ROCHA. 2006. An Urban Grammar and for the Medina Marrakech. *Design Computing and Cognition '06*: 483-502.

DUARTE, J. 2007. *Personalizar a habitação em série: Uma Gramática Discursiva para as Casas da MALAGUEIRA do Siza.* Lisboa: FCG-FCT.

GASPAR, J. 1969. A morfologia urbana de padrão geométrico na Idade Média. *Finisterra* IV, 8: 198-215.

HOLANDA, S. 1992. *Raízes do Brasil* (1936). Rio de Janeiro: José Olympio.

HORTA-CORREIA, J. 1997. *Vila Real de Santo António. Urbanismo e Poder na Política Pombalina.* Porto: FAUP.

Kappraff, J. 2002. *Beyond Measure.* Singapore: World Scientific.

Kruger, M. and Silva, C.2000. A Gramática da Forma das igrejas cistercienses. *Actas do Colóquio Cister: Espaços, Territórios, Paisagens.* Lisboa: 309-335.

KNIGHT, T. 1981. Languages of design: from Known to new. *Environment and Planning B:Planning and Design* **8**: 213-238.

———. 1989. Mughul Gardens Revisited. *Environment and Planning B:Planning and Design* **17**: 73-84.

LAMAS, J. 1993. *Morfologia Urbana e Desenho da Cidade.* FCG-JNICT.

LIEW, H. 2004. SGML: A Meta-Language for Shape Grammar. Ph.D. dissertation, Massachusetts Institute of Technology.

MARCH, L. 1976. *The Architecture of Form.* Cambridge: Cambridge University Press.

MARCH, L. and G. STINY. 1985. *Spatial systems in architecture and design: some history and logic.* Environment and Planning B: Planning and Design **12**: 31-53.

MARSHALL, S. 2009. *Cities, Design & Evolution.* London:Routledge.

MAYER, R. 2003. *A linguagem de Oscar Niemeyer.* MSc. Dissertation. Federal University of Rio Grande do Sul.

MITCHELL, W. J. 1998. The Logic of Architecture. Cambridge, MA: MIT Press.

MOREIRA, R. 1982. *Um tratado português de arquitectura do século XVI.* Master dissertation, Faculdade de Ciências Sociais e Humanas da Universidade Nova de Lisboa.

MURTEIRA, H. 1999. *Lisboa da Restauração às Luzes.* Lisboa: Ed. Presença.

OXMAN, R. 1997. Design by re-representation: a model of visual reasoning in design. *Design Studies* **18**: 329-347.

PAIO, A. 2007. Knowledge of geometrical design and composition in a Portuguese approach to urban layout. *Proceedings of the ISUF XIV International Seminar on Urban Form, XIV*: 212 - 232.

PAIO, A. 2009. Geometry, the Measure of the World. *Nexus Network Journal* **11**, 1: 63-76.

PAIO, A. and B. TURKIENICZ. 2009. A generative urban grammar for Portuguese colonial cities, during the sixteenth to eighteenth centuries. Towards a tool for urban design. *Proceedings of the eCAADe27*: 585-592.

———. 2010.A Grammar for Portuguese Historical Urban Design. *Proceedings of the eCAADe28. Future Cities*: 349- 358.

PIMENTEL, L.S. 1993. *Método Lusitânico de Desenhar as Fortificações das Praças Regulares e Irregulares* (1680). Lisboa: Direcção do serviço de fortificações e obras do exército.

REIS FILHO, N. 1962. *Contribuição ao Estudo da Evolução Urbana do Brasil (1500/1720)*. São Paulo: Pioneira Editora.

RHODEN, L. 1999. *Urbanismo no Rio Grande do Sul: Origens e Evolução*. Porto Alegre: Edipucrs.

RIBEIRO, O. 1962. *Portugal, o Mediterrâneo e o Atlântico*. Coimbra: Coimbra Editora.

ROSSA, W, ARAUJO, R., and CARITA, H. 1998. *Actas do Colóquio Internacional. Universo Urbanístico Português 1415-1822*. Lisboa: Comissão Nacional para as Comemorações dos Descobrimentos Portugueses.

ROSSA, W. 2002. *A Urbe e o Traço, Uma década de estudos sobre o urbanismo português*. Coimbra: Almedina.

ROSSA, W. and Luisa TRINDADE. 2006. Problems and precedents of the 'Portuguese city':understanding medieval urbanism and it's morphology. *Murphy. Journal of architecture history and theory* 1: 70-109.

SALGUEIRO, T. 1992. *A Cidade em Portugal. Uma Geografia Urbana*. Porto: Afrontamento.

SANTOS, P. 1968. Formação de cidades no Brasil colonial. *Actas do V Colóquio Internacional de Estudos Luso-Brasileiros*: 5-114.

SCHOLFIELD, P. H. 1958. *The Theory of Proportion in Architecture*. Cambridge: Cambridge University Press.

SMITH, R. 1955. Urbanismo colonial no Brasil. *Anais do II Colóquio Internacional de Estudos Luso-Brasileiros*: 1-42.

STEINØ, N. and N. E. VEIRUM. 2005. A Parametric Approach to Urban Design, Digital Design: The Quest for New Paradigms. *Proceedings of the eCAADe Conference*, Lisboa: 585 - 592.

STINY, G. 1978. *Algorithmic aesthetics. Computer Models for Criticism and Design in Arts*. Berkeley: University of California Press.

———. 1980. Introduction to shape and shape grammars. *Environment and Planning B: Planning and Design* 7: 343-351.

STINY, G. and J. GIPS. 1972. Shape grammars and the generative specification of painting and sculpture. *Information Processing* :1460-1465.

STINY, G. and W. J. MITCHELL. 1978. The Palladian Grammar. *Environment and Planning B: Planning and Design* 5: 5-18.

———.1980. The grammar of paradise: on the generation of Mughal gardens. *Environment and Planning B: Planning and Design* 7: 209-226.

TEIXEIRA, C. 1996. Portuguese Colonial Settlements of the 15th-18th Centuries. Vernacular and Erudite Models of Urban Structure. *La Ville Européenne comme Modèle*, Paris :15-26.

TEIXEIRA, M. and M. VALLA. 1999. *O Urbanismo Português, séculos XVIII-XVIII. Portugal Brasil*. Lisboa: Livros Horizonte.

TEIXEIRA, M. 2008. O Estado da Arte da Investigação Urbana em Portugal: A investigação dos núcleos urbanos de origem Portuguesa no mundo. *Revistas Ceurban* 8: 1-21.

TURKIENICZ, B. and R. MAYER. 2005. Cognitive Processes, Styles and Grammars. The cognitive approach applied to Niemeyer's free form. *Proceedings of the eCAADe23*: 529-536.

———.2006. Oscar Niemeyer Curves Lines: Few Words Many Sentences. Pp. 135-148 in *Nexus VI Architecture and Mathematics*, Sylvie Duvernoy and Orietta Pedemonte, eds. Torino: Kim Williams Books, 2006.

TURKIENICZ, B., E. WESTPHAL, and M. CAVALHEIRO. 2007. The Affordance between Context and Function as a means to stimulate the cognitive process in architectural design. *Shaping Design Teaching*. Aalborg: Aalborg University.

About the authors

Alexandra Paio is an architect with a Master's degree in urban design from ISCTE-IUL, Lisbon, Portugal. Currently she is a teaching assistant lecturing in Geometry and Architectural Composition at the Department of Architecture and Urbanism, ISCTE-IUL. Since 2007 she has been working on her Ph.D. thesis, entitled "Geometric models of representation in Portuguese urban design during the XV-XVIII centuries," at the same university. She has published papers in national and international journals and conferences about the relationship between geometry, architecture and urban design. She has participated in several workshops including the Shape Grammar Workshop by George Stiny (MIT), Faculdade de Arquitectura-UTL, Lisbon. She is a researcher in ADETTI/ISCTE.IUL.

Benamy Turkienicz holds a degree in architecture (Universidade Federal do Rio Grande do Sul, 1976), an MA in Urban Design (Joint Center for Urban Design - Oxford Polytechnic, UK, 1979), a MSc (Bartlett School of Architecture/University College London, UK, 1981), and a Ph.D. in Urbanism (Chalmers University of Tecnology, Sweden, 1982). He is a full professor at the UFRGS, teaching in the Department of Architecture of the Faculty of Architecture, the Program for Post-Graduation and Research in Architecture and the Program for Post-Graduation and Research in Design. He is an invited professor at the Master in Urbanism Erasmus Mundus and coordinates a Master Course in Planning Design at the Cape Verde University (cooperation protocol between CAPES/UniCV). He is co-author of several software applications to support architectural and urban design, including CityZoom. Turkienicz has delivered several lectures and conferences in universities in Brazil and abroad, including Nexus 2006 in Genoa, Italy. Currently, he is responsible for Post Graduate evaluation and accreditation in Architecture, Urbanism and Design at Capes, Ministry of Education in Brazil. Since 1995 he has been responsible for the SimmLab – Laboratory for the Simulation and Modeling in Architecture and Urbanism, Universidade Federal do Rio Grande do Sul, Brazil and for the NTU – Nucleo de Tecnologia Urbana/UFRGS, a joint research group which support municipalities and consulting firms in the planning and assessment of settlements and large scale architectural and urban projects.

Mário Krüger

Departamento de Arquitectura,
Faculdade de Ciências e Tecnologia
Universidade de Coimbra
3020-177 Coimbra, PORTUGAL
kruger@ci.uc.pt

José Pinto Duarte*
*Corresponding author

Faculdade de Arquitectura
Universidade Técnica de Lisboa
Rua Sá Nogueira
Pólo Universitário, Alto da Ajuda
1349-055 Lisbon PORTUGAL
jduarte@fa.utl.pt

Filipe Coutinho

Departamento de Arquitectura,
Faculdade de Ciências e Tecnologia
Universidade de Coimbra
3020-177 Coimbra, PORTUGAL
filipecoutinho@hotmail.com

Research

Decoding De re aedificatoria: Using Grammars to Trace Alberti's Influence on Portuguese Classical Architecture

Presented as a poster at Nexus 2010: Relationships Between Architecture and Mathematics, Porto, 13-15 June 2010.

Abstract. The research described in this paper is part of a project aimed at decoding Alberti's *De re aedificatoria* by inferring the corresponding shape grammar using the computational framework provided by description and shape grammars to determine the extent of such an influence in the Counter-Reformation period in Portugal. Here we concentrate on the theoretical foundations that enable the translation of Alberti's text into a shape grammar and then use it in determining its influence on Portuguese Renaissance architecture.

Keywords: theory of architecture, shape grammars, Leon Battista Alberti, De re aedificatoria, design automation; rapid prototyping

Introduction

This research project aims to understand the cultural impact of Alberti's *De re aedificatoria* by using an intelligent computational environment in order to grasp the full implications of that treatise for the architectural practice of classical architecture in Portugal and its overseas territories. In fact, an underlying influence of Alberti's theories is felt in classical architecture in Portugal, but no one has been able to determine the extent of such an influence. The idea is to translate the treatise into a description grammar [Stiny 1981] and a shape grammar [Stiny and Gips 1972] and then trace the influence of Alberti's work by determining the extent to which this grammar can account for the generation of Portuguese classical buildings. This approach follows the "transformations in design" framework proposed by Knight [1983], according to which the transformation of one style into another can be explained by changes in the grammar underlying the first style into the grammar of the second. Grammars are thus proposed as a complementary tool to be used by architectural historians to test hypotheses raised by documentary evidence. In the course of the project, the developed grammars will be implemented as computer programs, and the output digital models will be used to produce drawings, rapid prototyping models and virtual reality models. These elements will form part of an exhibition to be mounted at the end to describe and celebrate Alberti's work and its influence on Portuguese architecture.

State of the art

A controversial problem exists since there are authors who deny the existence of a Portuguese Renaissance architecture. Actually, Reynaldo dos Santos [1968-1970: II, 175] suggests that the Renaissance is a foreign style which had no influence on the

---

DOI 10.1007/s00004-011-0060-x; *published online* 4 February 2011

development of Portuguese architecture, and Jorge Pais da Silva [1986: 109] makes a direct transition between the Manueline and Mannerist styles without acknowledging the Renaissance period. However, more recently, Moreira [1991; 1995], in studying royal commissions between the Manueline style and the Roman mode of building, was able to identify one hundred and fifty buildings that can be considered as belonging to the Renaissance ambit.

With the help of a proper computational environment this project aims to clarify these issues and to develop a better understanding of Renaissance architecture in Portugal. Several questions can be raised, but the most important is to grasp how an emergent technology such as the construction of a shape grammar can help to spell out an issue that is deeply historically rooted.

The fact is that Alberti's treatise can be thought of as the production of algorithms with the aim of constructing intelligible architectural principles. Let us look, for instance, at the *columnatio* system as proposed in Book, VII, chap. 7 of the treatise, specifically, on the proportions of the Ionic base (see [Alberti 1988a: 203-204]).

In that case, all the proportional prescriptions of the Ionic base can be transformed into a computer program; in fact, all the issues raised by the dimensions and disposition of architectural elements, such as the column's diameter and its ratios, as well as the position of the torus, the die, the two scotias and rings, can be related within a computational framework in the sense that they are well defined problems and therefore can be spelled out as a set of forms and rules of transformation. Even the Ionic base can be thought of as a consequence of an antecedent form, the Doric base. The same happens with buildings, where, for instance, the sacred buildings described in Book VII are derived from the Roman basilica. In short, antecedent and consequent forms are deeply interrelated within the theoretical framework set out by Alberti.

Figure 1 – Outputs of a computer program developed after Vitruvius's *Ten Books of Architecture* using Autolisp, the scripting language of Autocad. From left to right and from top to bottom: digital models of theatres for 5,000, 10,000 and 15,000 spectators; and a 3D print model of the latter (Tiago Sousa and Duarte Pape, Programming and Fabrication for Architecture course, TU Lisbon, 2002. Instructor: J. P. Duarte)

These can be explored and developed to translate the rules set out in the treatise as a form of computation, as initially thought by Stiny [1980] and developed more recently, in the case of Siza's houses at Malagueira in Portugal, by Duarte [2001, 2005], using shape grammars to customize mass housing. Shape grammars have already been developed to encode algorithms codified in architectural treatises, for instance, by Andrew Li [2002], who developed a shape grammar after the rules described in a Chinese treatise from the twelfth century. Furthermore, computer programs can be developed from algorithms laid out in classical architectural treatises, as preliminary work developed in the context of a programming class for architects has shown (fig. 1). In this class, students translated the chapter on how to design Roman theaters from Vitruvius's *Ten Books of Architecture* into a computer program, then used this program to generated a digital 3D model of a theater for a given number of spectators, and finally used this model to produce a physical one using rapid prototyping. This work encompassed an intermediate step to translate the verbal instructions set in the treatise into a parametric model before writing the computer program in Autolisp, the scripting language of Autocad. A similar methodology will be used in the current project for translating Alberti's instructions for designing classical buildings into a computer program and then producing digital and physical models. However, the computational formalism used to translate verbal descriptions into a computer program is a grammar, not a parametric model. The reason for this choice is twofold. The first is a grammar's increased ability to encode complex instructions due to the possibility of combining different description and shape grammars into compound grammars; the second reason is its capacity to describe how to generate a design by sequential rule application, an important feature, given the didactic and pedagogic goals of the project and the aim of showing to which extent Alberti's rules were followed in subsequent Portuguese architecture.

Following this framework, Renaissance architecture in Portugal and in its overseas territories of Brazil and India can be thought as a sort of customization of the rules set out in the treatise by Alberti. That is our fundamental hypothesis for developing a shape grammar of Renaissance architecture in Portugal from the grammar of the original treatise. Work will proceed in five basic stages:

1. Inferring a shape grammar from the text of the treatise following a procedure similar to Li's [2002];

2. Structuring, testing and implementing this grammar with the buildings designed by Alberti as done by Duarte [2005];

3. Acknowledging transformations from the grammar of treatise to the grammar of the buildings designed by Alberti, as is identified by Krüger [2005] and in the notes by Krüger in [Alberti 2011] in a similar mode to that developed by Knight [1994];

4. Understanding the transformations from these sets of rules and forms in order to produce Renaissance architecture in Portugal and in the overseas territories, as is acknowledged by Carita [2008];

5. Implementing an educational software for generating and fabricating designs linking shape grammars and rapid prototyping following the model proposed by Duarte and Wang [2002];

6. Implementing an educational software, with an accompanying exhibition, in order to disseminate widely the results obtained so far.

Finally, a note on the final result. As there has already been an excellent exhibition of Alberti's works (see the catalogue of the exhibit at Palazzo Te in Mantua [Rykwert and Engel 1994]) using 3D models in wood as well as the CAD implementation of Alberti's buildings, our project can be thought of as a further development of this initial research with the aim of making Alberti's rules operational, but now as a form of computation to understand the architectural transformations that occurred in Portugal during the Counter-Reformation. In that sense, tradition and innovation are united by the idea that computation may play a major role in understanding Renaissance architecture in Portugal and, therefore, in shedding some light on the controversial issues mentioned above, raised by Santos [1968-1970], Silva [1986] and Moreira [1995].

Methodology

Alberti's treatise is one of the most influential architectural treatises of the Renaissance. Its translation into Portuguese was commissioned by King João III from André de Resende in the sixteenth century but was lost; its influence on Portuguese architecture remains elusive and is the subject of debate among scholars. The recent translation of Alberti's treatise from Latin into Portuguese [Alberti 2011] will provide the basis for determining the extent of such an influence and shed new light on the debate.

The aim of the project is, therefore, to trace the cultural influence of Alberti's treatise on Portuguese classical architecture, and the idea is to use the computational framework provided by description [Stiny 1981] and shape grammars [Stiny and Gips 1972] to determine the extent of such an influence in the Counter-Reformation period. To accomplish this, the following tasks will be necessary:

a. decoding the treatise by inferring the corresponding shape grammar;

b. comparing the grammar of the treatise with the buildings designed by Alberti;

c. tracing the influence of the treatise on Portuguese architecture by mapping the grammar of a particular building type in Portugal and its overseas territories to transformations of the initial grammar for the same type;

d. tracing the impacts of the treatise on the theory, practice and teaching of architecture by mapping the grammars underlying other theoretical and built works to transformations of the initial grammar;

e. disseminating the research results among scholars and the public in general by mounting a visually appealing exhibition using digital media.

A grammar consists of a set of substitution rules that apply recursively to an initial assertion to produce a final statement. In description grammars, the assertions are symbolic descriptions, whereas in shape grammars, they consist of shape descriptions.

The relation between description and shape grammars is such that for each shape rule there is a corresponding description rule and it is therefore possible to translate a description grammar into a shape grammar. Stiny and Mitchell [1978] have pointed out that a grammar can describe the formal and functional structure of a particular architectural style (descriptive value), explain how to synthesize new instances in the style (synthetic value), and determine whether a new instance is in the same style (analytical value). We will refer to these values in developing our research.

Alberti's treatise can be thought of as a set of algorithms that explain how to design buildings according to the canons of classical architecture. The aim is to translate the algorithms in the treatise into a description grammar, then into a shape grammar, and finally into a computer program. This is expected to be accomplished with the following four tasks:

1. Task 1, "understanding the treatise," aims at gaining a deep understanding of the treatise by manually designing buildings according to its rules and then producing the corresponding 3D models using rapid prototyping;

2. Task 2, "inferring the grammar," aims at developing the grammar, thereby gaining insight into the formal structure of Alberti's interpretation of classical architecture;

3. Task 3, "implementing the grammar," aims at writing the computer program encoding the grammar; and

4. Task 4, "matching the treatise grammar," aims at matching the grammar with actual buildings designed by Alberti and then with Portuguese classical buildings to determine the extent to which they match.

Knight [1983] has demonstrated that the transition of one particular style into a different but related one can be explained as a transformation of the grammar underlying the first style into that underlying the second by adding, subtracting, or transforming rules. The objective is to trace the influence of Alberti's treatise on Portuguese Counter-Reformation architecture by determining the extent to which the generation of buildings in this period might be explained by a transformation of the treatise grammar. Work in this stage will concentrate on one building type, possibly churches because they represent the most widely built and studied functional program, which means that there is enough scholarly work to make the current project feasible.

The traditional cultural approach to architectural history relies on the study of documentary sources to trace influences among theoretical and built works in the course of history. This approach is limited when there is not sufficient documentary evidence. An alternative approach is to study the inherent properties or architectural artefacts to determine the similarities and differences between them. This is particularly important in the case of functional and spatial properties which are often overlooked by historians mainly due to the lack of a rigorous methodology, but shape grammars can provide the required technical apparatus. This project aims to bring these two approaches together: namely, to use a cultural approach to trace possible influences of Alberti's treatise on the architecture that followed and then to use grammars to confirm such an influence, focusing on functional and spatial aspects. In addition, it is expected that this effort may constitute a basis for an enquiry into the usefulness of grammars and computational tools for the teaching and practice of architecture today. The aim of this project, therefore, is to trace the impacts of Alberti's treatise on the theory, practice and teaching of architecture by combining cultural and computational approaches.

Furthermore, Alberti recognizes that combinations of architectural elements alone would be meaningless, but joined together they can produce something well-designed, graceful and convenient:

> I am accustomed, most of all at night, when the agitation of my soul fills me with cares, and I seek relief from these bitter worries and sad thoughts, to think about and construct in my mind some unheard-of machine to move and carry weights, making it possible to create great and wonderful

things. And sometimes it happens that I not only calm the agitation of my soul, but invent something excellent and worthy of being remembered. And at other times, instead of pursuing these kinds of thoughts, I compose in my mind and construct some well-designed building, arranging various orders and numbers of columns with diverse capitals and unusual bases, and linking these with cornices and marble plaques which give the whole convenience, and a new grace [Alberti 1988b: III, 114-115].

This strongly suggests that the Albertian approach to generating architectural forms is capable of being developed by a shape grammar, which supports our aim of developing such a grammar within a Portuguese Counter-Reformation context.

Task 1: Understanding the treatise

This is a preparatory task required for gaining a deep understanding of the treatise. It has three purposes. The first purpose is to make an overall reading of its ten books (or chapters) in order to describe all the features that are capable of having an explicit visual grammar content. The second is to transform these explicit instructions into visual aspects of a grammar that will underlie the construction of a shape grammar such as:

a)   to make a clear cut distinction between lineaments, materials and construction;

b)   to describe the main differences between public and private buildings;

c)   to discriminate between the ornament of sacred, secular and private buildings.

The third purpose is to establish the main guidelines for visually developing Alberti's *columnatio* system, as well as the overall design strategies in public and private buildings and also in sacred and profane ones.

The aim is to delineate the main lines of inquiry that it is possible to develop in order to transform all the treatise lineaments into algorithms capable of providing an explicit configuration of Alberti's ideas.

This task will be developed in two phases, or subtasks. The first will involve all the researchers in the study of and reflection about the treatise. The second will create drawing models of the architectural building types.

Studying the original treatise and its translations into Portuguese and English

The seminal inspiration for this work was the translation of Alberti's treatise into Portuguese. As such, most of the work of gathering information has already been done. So, this initial task will simply focus on the study of both the original treatise in Latin, the plates later developed to illustrate the English translation, based on the 1550 Florentine edition of the treatise by Cosimo Bartoli, and the Portuguese edition by Mário Krüger and Arnaldo Espírito Santo, to be published in 2011 by the Calouste Gulbenkian Foundation [Alberti 2011].

Creating 2D and 3D models of the architectural artefacts described in the treatise

An effective way to understand the treatise is to create 2D drawings and 3D models of the described objects, including the column system and whole buildings, by applying the algorithms it describes "by hand". Such an understanding should help to clarify the algorithms and prepare the information for writing the grammars in the next tasks.

This will enable a deep understanding of the treatise and its translation required for supporting the subsequent work. Secondly, it will generate a set of algorithms and produce visualization material, including 2D drawings and 3D models, to be worked on in the next tasks.

Following this procedure a set of algorithms will be extracted that will serve as the basis for the subsequent development of the grammars in task 2. Drawings and models will serve as the basis for the development of the shape grammar by providing visualization material which will facilitate the analysis and comparison of buildings by Alberti with those designed and built within the Portuguese empire.

Task 2: Inferring the grammars

Grammar corresponding to the orders (*columnatio*)

The treatise describes the algorithms that should be followed in the design of buildings according to Alberti's interpretation of classical architecture. This task consists of the translation of the text of Alberti's treatise into shape [Stiny 1980] and description grammars [Stiny 1981], which requires the algorithms' rules to be codified into such grammatical formats.

The research will be concerned, firstly, with the development of a grammar for the column system and secondly with one for a chosen building type.

The classical orders establish a complex system of relations between the various parts and the whole of buildings according to a predetermined system of proportions.

The various parts that compose the column system (*columnatio*) are tokens of the larger formal language. This task will attempt to uncover the relationships between the Latin language, the original language of the treatise, and the way Alberti defines the canons of his interpretation of classical orders in Book VII. The idea is to write the text in a description grammar format and then write the corresponding shape grammar, thereby determining the algorithms underlying Alberti's treatise.

Grammar corresponding to a particular building type

The treatise also establishes very clear rules on how to design whole buildings, especially those described in Book VII (ornament to sacred buildings), Book VIII (ornament to public secular buildings), and Book IX (ornament to private buildings). This research will consider a given building type, such as churches, as a case study and then develop the corresponding grammars using the same methodology mentioned above. The grammar should codify the rules for generating both plans and elevations according to the proportioning system defined by the column system (*columnatio*).

Writing the rules contained in the treatise in grammatical format will systematize them in such a way that it will be possible to check whether Alberti followed such rules in the buildings he designed and will facilitate the development of the computer implementations.

These grammars will be used in Task 3 for comparing the rules Alberti laid down in the treatise with those he followed in the design of actual buildings and in Task 4 for developing the computer implementation.

Development of 2D and 3D digital representations

The goal is to build detailed 2D and 3D digital representations starting both from the output of the interpreters and from other sources of information, such as existing illustrations of the treatise, and from existing plans and elevations of Alberti's work and of Portuguese buildings influenced by it. This task encompasses the development of drawings, geometric models, texture mapping, scene creation and renderings. Some of the digital representations will be produced manually using the computer, whereas others will be produced as output of the various programs developed in the course of the project. Both types of digital models will be used to produce physical models using rapid prototyping.

Prototyping techniques

The development of grammars will uncover the rules encoded in the treatise and the computer implementation will permit their exploration. In Book I (*Lineamenta*), Alberti refers to 2D drawings and 3D models as the descriptive elements of building design. The project will therefore use various computer-based technologies for producing these types of elements using cutting-edge techniques. The use of rapid prototyping will enable a more tangible visualization of the output of the interpreters and particularly of Alberti's interpretation of classical architecture, thereby facilitating the fine-tuning of the grammars and increasing the didactic and pedagogical impacts of the research output.

Selection of rapid prototyping techniques

The purpose of this selection is to experiment with and select appropriate rapid prototyping techniques for producing physical models of the output of the grammar interpreters. Some of the techniques tested and used for the project are already available at partner institutions (laser cutting, 3D printing, CNC milling, etc.) and some will be acquired within the context of this project to increase the fabrication capabilities during and beyond the current project.

Production of physical models using rapid prototyping techniques

The objective is to study and illustrate the spatial and formal qualities of Alberti's interpretation of classical architecture and its influence on Portuguese architecture in the Counter-Reformation period. We hope to obtain a better understanding of the complexity of Alberti's thought on implementing and designing sacred and profane buildings, namely on the relationship between discursive and non-discursive thought. That is evident when Alberti invents new terms to describe elements of the column system, such as *rudens* (*rudentura* in Portuguese, "cable" in English).

The planned outcome of this task is a set of digital and physical models to be included in the exhibition at the end. These models are necessary first to gain a better understanding of Alberti's treatise and its impact on Portuguese architecture, and then to produce visually appealing material to include in the final exhibition, thereby achieving the didactic and pedagogical goals of the project. This task will result in a Master's thesis.

Task 3: Implementing the grammars

The codification of the treatise into grammars enables the complete generation of column systems and buildings according to the rules established by Alberti. The implementation of the grammars into computer programs will make such generation more efficient and enable the interactive exploration of the space of design solutions defined by the treatise in real time.

Selection of the computer environment

It will be necessary to choose the computer platform to use in the implementation of the grammars, that is, in the development of the grammar interpreters, including the choice of implementation paradigms and programming language. Based on previous experience in the development of interpreters for similar grammars, the interpreter will probably not need to support emergence, which will make implementation easier and faster to develop. Some of the languages that will be considered are Java, AutoLisp, RhinoScrip, VisualBasic, MaxScript, and Mel script. Among the criteria to be used to choose the language are the type and complexity of the forms to be generated, such as the geometrical intricacy of the Corinthian capital in Book VII, as well as whether to develop a standalone application or a macro within an existing CAD application.

Implementing the grammar of the column system orders

This task will concentrate on the implementation of the interpreter for the column system orders. The implementation will need to take into consideration issues of interface design to highlight the relations between the shape and description grammars and the complex system of formal relations that they define.

Implementing the grammar of the chosen building type

This task will focus on the development of the interpreter for the particular building type grammar. As with the column system grammar interpreter, this implementation will need to consider issues of interface design to emphasize and uncover the spatial relations between the elevation and the plan, as well as between the various parts of each 2D representation.

The result of these tasks will be two computer programs, one implementing the column system grammar and the other a selected building type. These programs will make it possible to test and refine the grammars and explore the universe of solutions defined by the encoded algorithms. They will also be part of the educational software to be implemented, which will focus on the development of the interface for the various programs produced in the course of the project so that their educational capabilities can be enhanced.

Task 4: Matching the treatise grammars

Although the treatise establishes very clear rules in written format, the lack of illustrations in the *editio princeps* highlights Alberti's intention to leave room for interpretation when designing real buildings and not jeopardize the accuracy of translating images from one manuscript to another. This task will study and compare built examples designed by Alberti with the treatise's rules in order to identify possible deviations from the canon and attempt to posit hypotheses to explain the source of such deviations.

The aim is to study and compare actual buildings designed by Alberti, namely S. Sebastiano and S. Andrea in Mantua, Malatesta's temple in Rimini, and the Palazzo Rucellai in Florence, with the treatise's rules in order to determine the extent to which the rules in the treatise were followed. This task will rely on Knight's "Transformations of Languages of Design" [1983] to explain the deviations rigorously as deletion, addition or transformation of the treatise's grammar rules.

The result will be a map with the eventual transformations of the treatise's grammar to account for the generation of the actual building by Alberti. The map should show which rules were deleted, added, or transformed, and it will serve as the basis to explain why such transformations were required in built examples. This map will then be compared with similar maps developed from the examples designed and built in each of the territories of the Portuguese empire studied. This comparison will help to shed some light on whether the transformations of the treatise within the studied Portuguese territory were due to a need to change the canons to respond to practical building constraints or to local cultural influences.

Mapping the transformations from the *De re aedificatoria*

The aim of this task is to map the transformations of the treatise grammar identified in the previous task to the geographical distribution of the selected case studies.

The idea is to be able to explain the geographical differences found in the same architectural type within that geographic area of the Portuguese empire as the successive transformation of the treatise grammar into local grammars and link those transformations to architectural, cultural, political, or other contextual influences.

In this task we will refer again to the framework proposed by Knight [1983] in Transformation in Design, in which the transformation of one style into another is explained by the deletion, addition, or transformation of rules in the initial grammar to obtain the final grammar.

The expected result is an extension of the map of the transformation of the treatise grammar into the grammar of Alberti's built works to encompass the transformation of the previous grammars into the grammars of churches in Portugal, Brazil and India in the studied period. This map is necessary to understand the extent of such transformations and thus of the influence of Alberti's treatise on the architecture of the Portuguese empire. It should also make it possible to formulate hypotheses regarding the source of those transformations, namely building constraints or cultural influences.

Conclusion

This paper describes the methodology to be followed in a research project whose aim is to determine the extent of Albert's influence on Portuguese classical architecture using a computational framework. Using this framework, the treatise will be translated into a shape grammar with the hope that by verifying the extent to which this grammar can account for the generation of Portuguese classical buildings, it will be possible to determine the extent of Alberti's influence. Specifically, by verifying the number of the treatise grammar rules that are used unchanged to derive a given building type (churches) it will be possible to determine how close the type is to Alberti's canons as set out in the treatise. Conversely, by discovering the number of rules that need to be changed, deleted from or added to the grammar, it will be possible to determine the degree of deviation of the Portuguese case from Alberti's principles. This methodology is proposed for use by architectural historians as a complement to the traditional use of documentary sources. The argument is that grammars might provide a rigorous method to test hypotheses raised by the use of such traditional sources.

Acknowledgments

The Digital Alberti project is funded by the Fundação para a Ciência e Tecnologia (FCT), Portugal (PTDC/AUR-AQI/108274/2008), hosted by CES at University of Coimbra, and coordinated by M. Krüger. F. Coutinho is also funded by FCT, grant SFRH/BD/ 66029/2009. We thank T. Knight, G. Stiny, and W.J. Mitchell for their support as consultants to the project. Sadly, W. J. (Bill) Mitchell passed away on June 11 2010. We are deeply thankful for his support to our research over the years. His insight and enthusiasm will be greatly missed.

Bibliography

ALBERTI, Leon Battista. 1966. *De Re Aedificatoria* (1485). 2 vols., Latin – Italian texts. Trans. Giovanni Orlandi, Intr. and notes by Paolo Portoghesi. Milan: Edizioni il Polifilo.

———. 1975. *De re aedificatoria. Editio princeps in facsimile.* Hans-Karl Lucke, ed. Munich: Prestel verlag.

———. 1988a. *On the Art of Building in Ten Books.* Joseph Rykwert, Neil Leach and Robert Tavernor, trans. Cambridge, MA: MIT Press.

———. 1988b. *Profugiorum ab aerumna libri III. Della tranquilità dell'animo.* (1441/1442) G. Ponte, ed. Genova: Casa Editrice Tilgher.

———. 2011. *Da Arte Edificatória.* Arnaldo Espírito Santo, trans; intr. and notes by Mário Júlio Teixeira Krüger. Lisbon: Calouste Gulbenkian Foundation (in print).

CARITA, Hélder. 2008. *Arquitectura Indo-Portuguesa na Região de Cochim e Keral.* Lisbon: Fundação Oriente/Transbooks.com.

DUARTE, José Pinto. 2001. Customizing Mass Housing: a discursive grammar for Siza's Malagueira houses. PhD Dissertation, Massachusetts Institute of Technology, Cambridge, MA.

———. 2005. Towards the mass customization of housing: the grammar of Siza's houses at Malagueira. *Environment and Planning B: Planning and Design* **32**, 3 (May 2005): 347-380.

DUARTE, José Pinto and Yufei WANG. 2002. Automatic Generation and Fabrication of Designs. *Automation in Construction* **11**, 3: 291-302.

KNIGHT, Terry W. 1983. *Transformations in design.* London: Cambridge University Press.

———. 1983. Transformations of language of design. *Environment and Planning B: Planning and Design* **10**, 2: (Part 1) 125-128; (Part 2) 129-154; (Part 3) 155-177.

———. 1994. *Transformations in Design: a Formal Approach to Stylistic Change and Innovation in the Visual Arts.* Cambridge: Cambridge University Press.

KRÜGER, Mario. 2005. *As leituras e a recepção do De Re Aedificatoria de Leon Battista Alberti.* http://homelessmonalisa.darq.uc.pt/MarioKruger/ParaumaLeituradoDeReAedificatoria.htm (Last accessed 15 November 2010)

LI, Andrew I-Kang. 2002. Algorithmic Architecture in Twelfth-Century China: the *Yingzao Fashi.* Pp. 141-150 in *Nexus IV: architecture and mathematics,* José Francisco Rodrigues and Kim Williams, eds. Fucecchio, Florence: Kim Williams Books.

MOREIRA, R. 1991. A Arquitectura do Renascimento no sul de Portugal. A encomenda régia entre o Moderno e o Romano. Ph.D. thesis. Lisboa: FCSH-UNL.

———. 1995. Arquitectura: Renascimento e Classicismo. Vol II, pp. 302-375 in *História da Arte Portuguesa,* P. Pereira, ed. Lisbon: Circulo dos Leitores.

RYKWERT, Joseph and Anne ENGEL. 1994. *Leon Battista Alberti.* Milan: Olivetti/Electa.

SANTOS, Reynaldo dos. 1968-1970. *Oito séculos de Arte Portuguesa.* 3 vols. s/d. Lisbon: Empresa Nacional de Publicidade.

SILVA, Jorge Henriques Pais da. 1986. *Paginas de História de Arte.* 2 vols. Lisbon: Ed. Estampa.

STINY, George. 1980. Introduction to Shape and Shape Grammars. *Environment and Planning B: Planning and Design* **7**, 3: 343-352.

———. 1981. A note on the description of designs. *Environment and Planning B: Planning and Design* **8**, 3: 257-267.

STINY, George and James GIPS. 1972. Shape Grammars and the Generative Specification of Painting and Sculpture. Pp. 1460-1465 in *Information Processing 71,* C.V. Freiman, ed. Amsterdam: North-Holland. Rpt. in *Best computer papers of 1971,* O. R. Petrocelli, ed. Philadelphia: Auerbach, pp. 125-135.

About the authors

Mário T. Krüger holds a Diploma in Architecture from the Lisbon School of Fine Arts (1972), a Master in Urban Science, University of Birmingham (1973), a Doctorate in Architecture from the University of Cambridge (1977), and a Post-Doc from the Bartlett School of Architecture, University College (1991). Currently he is Full Professor in the Department of Architecture at the University of Coimbra. He has taught at the Technical University of Lisbon School of Architecture, at the Instituto Superior Técnico, and at the University of Brasilia where he was Head of the Department of Urbanism. He has authored over 47 journal articles and 3 books. His most recent work is the annotated translation of Alberti's *De Re Aedificatoria*. His research interests include the study of Alberti's work and the impact of computation on architectural theory and practice.

José Pinto Duarte holds a B.Arch. (1987) in architecture from the Technical University of Lisbon and an S.M.Arch.S. (1993) and a Ph.D. (2001) in Design and Computation from MIT. He is currently Visiting Scientist at MIT, Associate Professor at the Technical University of Lisbon Faculty of Architecture, and a researcher at the Instituto Superior Técnico, where he founded the ISTAR Labs - IST Architecture Research Laboratories. He is the co-author of *Collaborative Design and Learning* (with J. Bento, M. Heitor and W. J. Mitchell, Praeger 2004), and *Personalizar a Habitação em Série: Uma Gramática Discursiva para as Casas da Malagueira* (Fundação Calouste Gulbenkian, 2007). He was awarded the Santander/TU Lisbon Prize for Outstanding Research in Architecture by the Technical University of Lisbon in 2008. His main research interests are mass customization with a special focus on housing, and the application of new technologies to architecture and urban design in general.

Filipe Coutinho holds a Bachelor of Architecture degree from the Technical University of Lisbon Faculty of Architecture (1994) and a Master of Science in Building Technology from the Instituto Superior Técnico (2004). He is currently a Ph.D. student at the University of Coimbra.

Andrew I-kang Li

Initia Senju Akebonocho 1313
Senju Akebonocho 40-1
Adachi-ku
Tokyo 120-0023, JAPAN
i@andrew.li

Keywords: style, shape
grammars, design, design theory

Research

Computing Style

Presented as a poster at Nexus 2010: Relationships Between
Architecture and Mathematics, Porto, 13-15 June 2010.

Abstract. Shape grammars are frequently used in analyzing
style in architecture and other areas of design. But this is a
more subtle task than is usually realized, and some
grammatical approaches to design analysis are logically
suspect. We examine the framework articulated by Stiny and
Mitchell in 1978, fill in the operational gaps, and propose a
more comprehensively considered approach to using
grammars to compute style.

Introduction

One of the most perceptive contributions to shape grammar studies is, strictly
speaking, not about shape grammar at all. Rather, it is about the more general problem of
style, what it means to understand style, and how computation helps to achieve that
understanding. In the conclusion of their Palladian grammar, Stiny and Mitchell [1978]
described the problem of style analysis as follows (in what follows we paraphrase and
generalize from buildings to designs).

We are given a finite set of designs that appear to be similar. This set, or *corpus,* is a
subset of a set of all and only those designs that are also similar (i.e., a *target language* or
style). Understanding this style means being able to do three things: first, create a new
design in the same style; second, evaluate whether a previously unknown design is in the
same style or not; and third, explain why the designs appear to be similar.

This is an inspired application of the scientific method to design analysis. It frames
the question clearly as one of observation and operation: *understanding* means *being able
to do.* This in turn explains, for example, why connoisseurship is ultimately
unsatisfactory as a demonstration of understanding: the connoisseur is by definition a
black box, unable or unwilling to explain how he does what he does.

All formal generative tools, such as programming languages and shape grammars,
provide these three capabilities. But shape grammars have the additional advantage of
being graphic, which makes them uniquely well-suited to studying style in design, which
employs graphic, not symbolic, artifacts.

For the first task, we create a new design in the style by using the grammar to
generate it. For the second, we use the grammar to try to generate the previously
unknown design; if we can, it is in the style, and if not, not. And for the third task, that
of explaining, the grammar embodies a generative algorithm: it is itself the explanation.[1]

At the same time, however, Stiny and Mitchell are silent on the very real differences
between design and natural phenomena. How, for example, are we to evaluate similarity?
Using a grammar to generate a design is uncontroversial, but how do we know whether
that design is in the style? You and I may not agree, and where would this leave us? This
seems to be a hopelessly subjective question.

In fact, the question *is* subjective, but this detracts not at all from Stiny and Mitchell's elegant statement. If we examine this question more closely, we will gain a richer appreciation, not only of how computation can help us understand style, but also of how we humans are ultimately central to the very idea of style. The formal and the informal, the objective and the subjective all have a place, and we can come to terms with them by wrestling with the paradoxes.

Style analysis and science

The hypothesis proposed by a natural scientist makes predictions, and she evaluates those predictions by doing experiments. These experiments are essentially questions from the scientist, and are designed so as not to distort the answers from nature. To put it another way, the scientist and nature must be independent. We will emphasize this by referring to the natural scientist as the *investigator* and to nature as the *evaluator*.

In other sciences, such as the humanistic sciences, the evaluators may be humans. In linguistics, for example, a linguist studying a language proposes a grammar that generates new sentences (i.e., ones not in the corpus). A *native speaker* of that language evaluates whether those new sentences are *legal* (i.e., in the language). The linguist and the native speaker are independent: the linguist does not evaluate the sentences, and the native speaker does not create the grammar. In addition, we need to point out that the native speaker is *authoritative:* his judgement, like nature's, is accepted by the investigator and, she hopes, her audience.

Returning from science to style, it is now easy to imagine a setup with an independent and authoritative human evaluator analogous to the native speaker; we will call this evaluator a *native stylist*. The style investigator proposes a grammar (i.e., a hypothesis) that generates new designs, and the native stylist judges whether or not the designs are legal (i.e., in that style).

As an example, let us examine the case of José Pinto Duarte, who studied the houses designed and built in Malagueira, Portugal, by the architect Álvaro Siza [Duarte 2005]. Duarte proposed a grammar (without Siza's input), generated new designs, and presented them to Siza to evaluate. Siza is the perfect native stylist: he is independent of Duarte, and he is clearly authoritative with respect to his own work.

What did Siza think of Duarte's designs? He had four basic responses, varying in degree of legality.[2]

1. "No, it's not legal, but the difference is small."
2. "Yes, it's legal, but I wouldn't have done it for idiosyncratic reasons."
3. "Yes, it's legal, and I might have designed it."
4. "Yes, it's legal, and I designed it. ... What do you mean I didn't design it?"

Overall, the four responses suggest that Duarte's grammar was quite accurate. But they are worth a closer look. Siza's first response suggests that the boundary is clear to him, and yet also that illegality comes in shades of gray. His second response suggests that he recognizes two standards simultaneously. Here it may be well to realize that his approach in Malagueira was to develop a method that his assistants used to create house designs. Thus the first standard is for his staff; the second is for himself. The design in question is legal according to the first standard, but illegal according to the second. Taken together, these two responses suggest that the boundary is flexible, and that even the perfect native stylist may not be entirely consistent.

The case of Duarte and Siza may seem charming, but much of the charm is simple novelty in a world where native stylists are rarely available when we need them. We may want to study churches by Hawksmoor, but Hawksmoor is not at hand to evaluate our predictions.

Without independent and authoritative evaluation, we cannot know whether the designs predicted by our grammar are legal or not. The only designs that we know to be legal are those in the corpus. But this we knew before we proposed the grammar; we have learned nothing. Scientific experiments add to the known facts, but without an appropriate mechanism for evaluation, our grammatical setup does not. The number of known facts matters, according to a principle that scientists follow when selecting from competing hypotheses. This principle, known as Ockham's Razor, guides them to select the most parsimonious hypothesis that accords with the known facts. Thus, all else being equal, the more known facts that a hypothesis accounts for, the more powerful it is. And since grammars with unevaluated predictions do not add to the known facts, we lose one criterion for selecting among competing grammars.

With this in mind, we see that, because Duarte gets a reliable verdict on his predictions, he has established more known facts with which all grammars by subsequent investigators must be consistent. By contrast, consider the case of you, me, and Hawksmoor. Suppose that I propose one grammar and you propose another. How are we to choose between them? Without Hawksmoor's participation, my predictions and yours are simply speculative. All we know for sure is what we started with: the churches in the corpus. It's my word against yours.

At this point we would like to mention the role of embedding in the situation we have just described. In particular we would like to note that it echos Stiny's [1980; 2006; 2011] consistent contentions that embedding is central to how we perceive the world and that shape grammars help us do that.

We have mentioned already that the corpus is a subset of the target language: it is *embedded* in the target language. At the same time, however, it is embedded in infinitely many other languages of designs. Our goal as investigators of style is to determine which of all these languages is the target language, even though, given only a corpus of designs, we cannot do this with any certainty. This condition evinces the inherent ambiguity of the inductive method.

Flawed approaches to grammatical style analysis

The framework that we have proposed has implications for how grammatical style analysis should be conducted. To begin, let us consider briefly some approaches seen in the literature. Some of these approaches are inconsistent with our framework; we believe that their results are therefore flawed.

In some cases, the investigator assumes that any design generated by her grammar is legal, because it was generated. She mistakes existence for legality. This is simply proof by assertion. To the extent that there is any evaluation at all, the investigator is acting as her own evaluator, which, as we have seen, is untenable.

In other cases, the investigator recruits lay people, such as students, exposes them to the corpus, and then uses them as evaluators. Having been trained by the investigator, they are not independent, and, being novices, they are not authoritative. As evaluators, they are not credible.

In still other cases, the investigator uses a connoisseur. A connoisseur is independent, but, like the native speaker and native stylist, is not always available. Unlike those informants, the connoisseur possesses an authority that is not innate but acquired. As a result, the quality of his claim to authority significantly determines the persuasiveness of the investigator's results.

Sometimes the investigator removes a small number of designs from the corpus, puts them aside, and proposes a grammar based on the remaining designs. If the grammar generates the removed designs, then she considers those predictions to be confirmed. This has the advantage of making an evaluator unnecessary for those designs, but has disadvantages.

For one, it is difficult for the investigator to disregard the removed designs while composing the grammar. This could taint the independence of the predictions and their confirmation. For another, the grammar makes many other predictions, which remain unevaluated. In fact, since the "confirmed" designs come from the corpus, this setup does not actually add to the known facts. After all, if those designs are left in the corpus, the grammar still has to generate them. So the investigator has not accomplished anything more than she would have otherwise.

A "purer" version of this approach is to discover a new and credible instance of the style defined by the grammar. This is a chancy approach to evaluation, because the investigator can never be sure of finding a new and credible design when she needs it. But let us consider what this procedure might look like.

Suppose that our investigator proposes a grammar of Palladian villas. If she then discovers a previously unknown but authentic edition of the *Four books,* and it contains a villa that was predicted by the grammar but was not part of the original corpus, then she can consider that design legal; one of her grammar's predictions is confirmed. In effect, the newly discovered edition is the evaluating mechanism.

However, this begs the question: how do we know that this edition is authentic? We must examine the book itself and the circumstances of its writing and publication, and consider them in the light of what is already known about Palladio's work, life, and times. In other words, we have to evaluate, not the design itself, but the new evaluating mechanism. This suggests a workable approach to evaluation, which we now discuss.

Our approach to grammatical style analysis

Our approach is this. Once we have our predictions, we imagine what information we would need to evaluate them, we look for that information, and then we use it to evaluate them. As a hypothetical example, suppose that we generate a new Hawksmoor church. Does this design make sense structurally? We consider the technological practise of Hawksmoor, his time, and his place. And then we evaluate the design against that background. Does the design make sense functionally? We consider the religious practise that Hawksmoor was accommodating in that time and place, and consider the design in that light. We can consider any aspect, any area that we deem relevant.

The questions that matter are not grammatical ones. In the end, it all comes down to *us,* not as the evaluator of the design itself, but as the finder, evaluator, and applier of evaluating mechanisms. The evaluation is *indirect.* It requires imagination, as Barzun and Graff remind historians: "The researcher must again and again *imagine* the kind of source he would like to have before he can find it" [1992, 58–59, original emphasis].

This is exactly what Liang Sicheng, the pioneering historian of Chinese architecture, did when he began to study the hitherto indecipherable twelfth-century Chinese government building manual *Yingzao fashi* (*Building standards*). Wanting to study extant buildings, and imagining that some must exist in the countryside, he surveyed historical documents and canvassed local people to identify buildings that might merit on-site investigation. This led to fieldwork and numerous important discoveries [Fairbank 1994]. We now turn to this book to illustrate more concretely just how much imagination and judgement are involved in grammatical style analysis.

An example

The *Yingzao fashi* [Liang 1983] was written by the court architect, Li Jie (d. 1110), and published in 1103 (Song dynasty, 960–1127).[3] This book of standards was intended to enhance communication between the government (as a client) and builders it employed, and contains mostly written guidelines and some drawings. Among the drawings are eighteen sections of a building type known as a *ting* hall (figs. 1-4); these form our corpus.

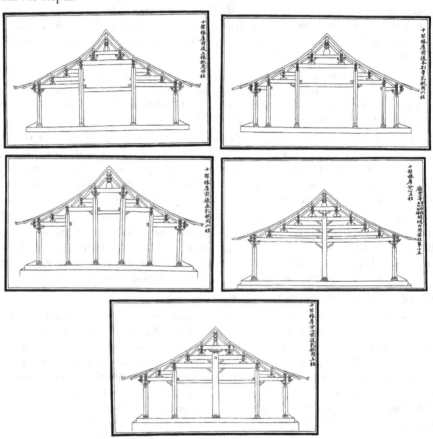

Fig. 1, a–e. Ten-rafter *ting* hall sections in the *Yingzao fashi* [Liang 1983: 313–15]

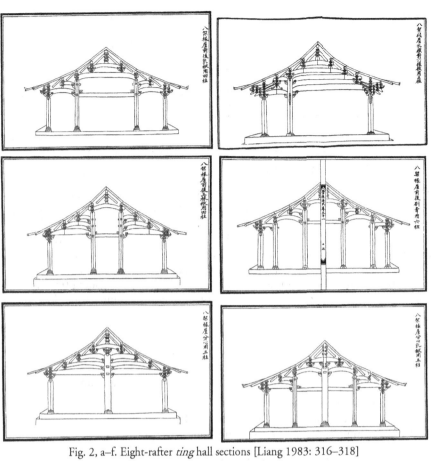

Fig. 2, a–f. Eight-rafter *ting* hall sections [Liang 1983: 316–318]

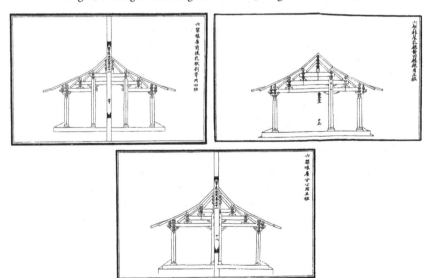

Fig. 3, a–c. Six-rafter *ting* hall sections [Liang 1983: 319–20]

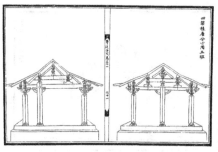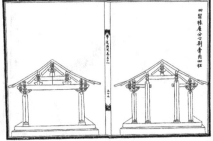

Fig. 4, a–b. Four-rafter *ting* hall sections (Liang 1983, 320–21).

Determining whether these drawings are reliable for our purpose is our first test of (extragrammatical) judgement. They appear at least 250 years more recent than the text, as the building components clearly reflect the style of the Ming and Qing dynasties (1368–1644, 1644–1911). It may be that Ming or Qing artists copied the drawings from an earlier edition (now lost), in the process substituting contemporary details. For our purpose, though, we can consider them simply diagrams, and judge that as such they probably preserve the information relevant to us: the depth of the building (measured by the number of segmented rafters in the roof structure) and the number and location of columns.

As we can see in the corpus, the building depth is 4, 6, 8, or 10 rafters; there is one column below the front eaves and another below the back eaves; and inside the building there may be additional columns below purlins (where adjacent rafters meet). It might seem obvious that we should infer the following guidelines for our grammar:

1. The number of rafters must be even.
2. The number of rafters must be not less than 4 and not more than 10.
3. Columns must be located below purlins (not between them).

However, we cannot be certain about any of these conclusions. For the number of rafters, a number such as 11 may be unlikely, but who is to say it is impossible? Perhaps an 11-rafter building once existed, but has since been lost. Perhaps it is a black swan, waiting to be discovered. In fact, if we consult extant buildings, we find that they overwhelmingly conform to the first two guidelines. We also see 2-rafter structures, which we might dismiss as being only gates and not buildings. In the Ming and Qing dynasties, we see buildings which appear to have an odd number of rafters, but we might consider the single curved rafter at the ridge as a fusion of two rafters. If so, we could dismiss these buildings as being irrelevant variants of even-numbered ones or simply anachronistic. Thus we might still keep the first two guidelines, but only after searching for and evaluating relevant evidence, and convincing ourselves.

The location of columns is trickier. Extant buildings with columns between purlins, not below them, actually do begin to appear about 150 years after the publication of the *Yingzao fashi*, in an area by then long lost from Song control (see fig. 5).[4] The question now is whether we should consider such buildings as examples of Song standards.

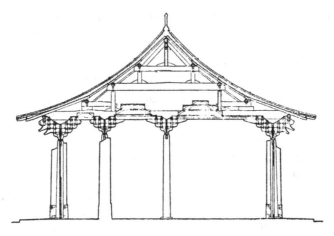

Fig. 5. The Sanqing dian of Yongle gong (1262), with columns located between purlins [Du 1963]

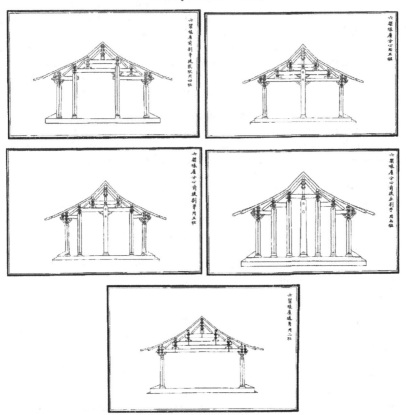

Fig. 6, a–e. New 6-rafter *ting* hall sections generated by our grammar (redrawn in the style of the *Yingzao fashi*)

We might survey extant buildings (and secondary evidence such as paintings and sculpture) and in that context conclude that columns between purlins represent a significant structural departure from previous practise. If so, we might argue that such buildings do not belong in the style defined by the *Yingzao fashi* and so should not be generated by our grammar. This would uphold the third guideline.

On the other hand, others might argue that the stylistic break at the end of the fourteenth century (between the Yuan and the Ming dynasties) is far greater (see [Liang 1984]), and that columns between purlins are simply variations within the style of the *Yingzao fashi*. They would then discard the third guideline. Again, it matters what evidence we examine and how we evaluate it; this requires historical judgement, not grammatical technique.

The guidelines we have discussed above are actually generic predictions that we evaluate *before* the grammar can generate them. Assuming that we have confirmed the three guidelines, we now have several predictions for 6-rafter sections, five of which we show in fig. 6.

Let us look at the fourth design, which has seven columns. How should we evaluate it? There is no such extant building, and we might invoke an architect's understanding of function to conclude that a building with no large spaces is unusable. Others might object that extant buildings are not government buildings, but temples, which require large spaces for gatherings. They would find the lack of built evidence irrelevant, and might instead search for a better understanding of how government buildings were used, perhaps in government documents. They would then head for the archives.

The fifth design has no interior columns; this arrangement has its own name: *tongyan,* or *clear span.* The lowermost beam, being undivided by interior columns, is 6 rafters long. In the corpus there is one clear span section, four rafters deep (fig. 4, 3rd section); there is no such section for depths of six, eight, or ten rafters. Should we take the 4-rafter clear span as representative and conclude that a 6-rafter clear span is legal? Or should we take the absence of 6-, 8-, and 10-rafter clear spans to be decisive? We might also note that a 6-rafter beam is seen in the corpus (fig. 2, 2nd section), which would suggest that 6-rafter-long beams were physically available and that 6-rafter clear span halls were structurally feasible. This might prompt us to study the absolute lengths of beams in extant buildings in various times and places to understand what materials were available, when, and where. (It is nice to notice that this is yet another example of embedding: the beam can be read as a subunit of the building or as a member with an absolute length, because it is both.)

This example shows the type of knowledge and judgement that go into the development of a mechanism to evaluate the predictions of a grammar. What is important here is that this knowledge and judgement have little to do with grammar and much to do with virtually anything else that we find relevant. By contrast, working out a grammar that produces designs that the mechanism finds legal is a relatively straightforward technical exercise. (Only relatively, of course: judgement is still called for in creating a grammar that explains the basis of the style clearly and persuasively.)

Conclusion

We have seen that, to evaluate design predictions credibly, we must exercise our judgement in matters that are unrelated to grammars. However, this is not cause for despair. On the contrary, the combination of grammars and Stiny and Mitchell's framework is a powerful tool with which we can confront the paradox of hypotheses without direct evaluation. Grammars make explicit where designs come from, but they also can help us become aware of how those designs are evaluated. We should be mindful of the operational subtleties in constructing an evaluation mechanism: what knowledge we bring to bear, what criteria we develop, how we apply them. Then we can say what we are doing, and we can debate differing views. Thinking critically and imaginatively in this way is the essence of humanistic study.

Our approach to evaluation is consistent with the views of both Frege and Wittgenstein, which are sometimes thought to be contradictory. In Frege's view:

> there must not be any object as regards which the definition leaves in doubt whether it falls under the concept; though for us human beings, with our defective knowledge, the question may not always be decidable. We may express this metaphorically as follows: the concept must have a sharp boundary [1977: 259].

A grammar defines a style with sharp boundaries, but the impossibility of equally sharp evaluation arises precisely because of our "defective knowledge." Wittgenstein [1958], on the other hand, insists on the possibility of vague boundaries. Each of our grammatical boundaries is sharp, but, because you and I may disagree or because, like Siza, you or I may even hold differing views simultaneously, they are also changeable and not mutually exclusive. Taken together, they are vague.

Style is not "out there," waiting to be discovered. Rather, it is constructed, both through the grammar and through the mechanism by which we evaluate our grammar's predictions. And it is the quality of this construction, more than the style itself, that should interest us. A definition of style is only as good as its evaluation mechanism. In this sense, to attend with care and self-awareness to both grammar and evaluation is to compute style.

Notes

1. This is not a new idea. As Dijkstra said: "I view a programming language primarily as a vehicle for the description of (potentially highly sophisticated) abstract mechanism" [1976: 9].
2. Paraphrased from an e-mail from Duarte dated 15 August 2003.
3. For an introduction in English, see [Glahn 1984]; see also [Li 2002].
4. This disposition is known as *yizhu* (shifted columns).

References

BARZUN, Jacques, and Henry F. GRAFF. 1992. *The modern researcher.* 5th ed. Fort Worth: Harcourt Brace Jovanovich.

DIJKSTRA, Edsger W. 1976. *A discipline of programming.* Englewood Cliffs, N.J.: Prentice-Hall.

DU Xianzhou. 1963. Yongle gong de jianzhu [The architecture of Yongle gong]. *Wenwu* 1963, no. 8: 3–17.

DUARTE, José Pinto. 2005. Towards the mass customization of housing: the grammar of Siza's houses at Malagueira. *Environment and planning B: Planning and Design* 32, 3: 347–380.

FAIRBANK, Wilma. 1994. *Liang and Lin: partners in exploring China's architectural past.* Philadelphia: University of Pennsylvania Press.

FREGE, Gottlob. 1997. Grundgesetze der Arithmetik, volume II. In *The Frege reader,* edited by Michael Beaney, 258–289. Oxford: Blackwell Publishers.

GLAHN, Else. 1984. Unfolding the Chinese building standards: research on the *Yingzao fashi.* Pp. 47–57 in *Chinese traditional architecture,* Nancy Shatzman Steinhardt, ed. New York: China Institute in America.

LI, Andrew I-kang. 2002. Algorithmic Architecture in Twelfth-Century China: the *Yingzao Fashi*. Pp. 141-150 in *Nexus IV: Architecture and Mathematics*, José Francisco Rodrigues and Kim Williams, eds. Fucecchio, Florence: Kim Williams Books.

LIANG Sicheng. 1983. *Yingzao fashi zhushi* [The annotated *Yingzao fashi*]. Beijing: Zhongguo jianzhu gongye.

LIANG Ssu-ch'eng. 1984. *A pictorial history of Chinese architecture: a study of the development of its structural system and the evolution of its types.* Wilma Fairbank, ed. Cambridge, MA: MIT Press.

STINY, George. 1980. Introduction to shape and shape grammars. *Environment and planning B: planning and design* 7: 343–351.

———. 2006. *Shape: talking about seeing and doing.* Cambridge, Mass.: MIT Press.

STINY, George, and William J. MITCHELL. 1978. The Palladian grammar. *Environment and planning B: Planning & Design* **5**, 1: 5–18.

WITTGENSTEIN, Ludwig. 1958. *Philosphical investigations.* 2nd ed. Oxford: Basil Blackwell.

About the author

Andrew I-kang Li is an independent researcher in computational design based in Tokyo. His work ranges from a computational analysis of the twelfth-century Chinese building manual *Yingzao fashi* to a software application for creating and editing shape grammars. He is presently an adjunct professor at Korea Advanced Institute of Science and Technology (KAIST) and president of the Association of Computer-Aided Architectural Design Research in Asia (CAADRIA). He was at the School of Architecture, The Chinese University of Hong Kong, from its founding in 1991 until 2010, and has also taught at Tunghai University, Taiwan, and worked as an architect in Boston, USA. Dr. Li was born in Montréal, Canada, and has an L.Mus. in piano performance (McGill University, Canada), an A.B. in Chinese (Harvard University, USA), an M.Arch. (Harvard), and a Ph.D. in computational design (Massachusetts Institute of Technology, USA). He studied Chinese architectural history at Nanjing Institute of Technology (now Southeast University) as a China / Canada government exchange scholar. He presented "Algorithmic Architecture in Twelfth-Century China: the Yingzao Fashi" at the Nexus 2002 conference (pp. 141-150 in Nexus IV: Architecture and Mathematics, eds. Kim Williams and Jose Francisco Rodrigues, Fucecchio (Florence): Kim Williams Books, 2002). See http://www.nexusjournal.com/conferences/N2002-Li.html. He is a member of the Editorial Board of the *Nexus Network Journal*.

Dirk Huylebrouck[*]

*Corresponding author

Department for Architecture
Sint-Lucas,
Paleizenstraat 65
1030 Brussels BELGIUM
huylebrouck@gmail.com

Antonia Redondo Buitrago

Departamento de Matemáticas.
I.E.S. Bachiller Sabuco,
Avenida de España 9
02002 Albacete SPAIN
aredondo@sabuco.com

Encarnación Reyes Iglesias

Departamento de Matemática
Aplicada.
Universidad de Valladolid
Avenida Salamanca, s/n
47014 Valladolid, SPAIN
ereyes@maf.uva.es

Research

Octagonal Geometry of the Cimborio in Burgos Cathedral

Abstract. This paper is a geometric analysis of elements of Burgos Cathedral, featuring the Cordovan proportion. It invites to mathematical and decorative creativity with rosettes and tesseracts, based on other remarkable proportions as well.

Keywords: Burgos Cathedral, geometric analysis, Cordovan proportion, regular polygons, geometric shapes, stars, tesseracts, silver mean

Fig. 1. Burgos Cathedral

Introduction

Burgos is a city in Spain in the region of Castilla and León (fig. 2). The stylized spires and stately towers of its cathedral are among the most beautiful achievements of Gothic art in Spain (fig. 1). Burgos Cathedral was declared a World Heritage Site by UNESCO [Rico Santamaria 1985].

Fig. 2. Castilla-León (grey) in Spain

This magnificent work of art was pieced together and erected above an ancient Romanesque cathedral of the eleventh century (begun in 1075 and completed in 1096), built by Alfonso VI, named Santa María, where the famous Spanish nobleman and military leader Rodrigo Díaz de Vivar, called Cid Campeador, departed for exile. This historic event and the heroic deeds of that Castilian knight are narrated in the Spanish

Nexus Network Journal 13 (2011) 195–203   NEXUS NETWORK JOURNAL – VOL.13, No. 1, 2011 **195**
DOI 10.1007/s00004-011-0057-5; *published online* 26 February 2011
© 2011 Kim Williams Books, Turin

epic poem "Cantar del Mío Cid", which mixes the history of the medieval wars between Spanish Christian Kingdoms and Arabs with legends.

The current cathedral is the result of the efforts and interventions of a collective: kings, bishops, architects, sculptors, artisans, marble workers, carvers, blacksmiths, etc. Many circumstances promoted the construction of the new cathedral at the beginning of the thirteenth century: a historic dynamism, the support of the city, the strategic situation of the town of Burgos in the Camino de Santiago, and, mainly, the appointment of bishop Mauricio, a cultured man trusted by King Ferdinand III the Saint. After travelling to France and Germany, the prelate urged Ferdinand King to construct a new cathedral, vast and stately. On 20 July 1221, the foundation stone of the new cathedral was laid.

The architectural solutions in Burgos Cathedral were influenced by French Gothic design and, after the fifteenth century, by Germanic artistic inspirations as well. Its construction went through many phases and styles: from classic Gothic, (from 1221 through the first half of the fourteenth century), to late Gothic, (from the second half of the fourteenth century through the first half of the sixteenth century). Some elements of Renaissance and Baroque styles are present as well, but they are less prominent.

Geometric patterns

Some significant elements of the cathedral were built in the second phase: the octagonal spires, the octagonal Condestable Chapel (a *condestable* held the highest rank in the military profession during the Middle Ages), and the most significant construction of that time, an octagonal tower in Gothic-Plateresque style, known as the cimborio. This tower, the highest in the Cathedral, corresponds to the lantern of the transept and is configured as an octagonal prism within a squared prism with four towers on the vertices of the latter. On the vertices of the octagonal prism also rise eight smaller towers. The octagonal tower is placed over the crossing of the nave where the tomb of Cid Campeador is located.

The geometry of the cimborio becomes more discernable when entering through the door of the Sarmental façade that opens to the transept of the cathedral. Then the tower can be observed from the inside, in its entire splendor. It is an octagonal prism supported by means of pendentives in another quadrangular prism, held up by four great columns (fig. 3). In the middle there is a well-designed rosette (fig. 4).

Fig. 3. Vault of the cimborio

Fig. 4. Pattern against the light (left); geometric pattern of the rosette (right)

The initiative for the building of the first tower came from Bishop Luis de Acuña, who had the idea for the construction of a lantern for the transept in 1460. The master builder was Juan de Colonia (Köln) and he incorporated the German building trends and methods of that time.

In 1481 Juan died, and his son, Simón de Colonia, continued the project, finishing the tower in 1489. However, the current cimborio doesn't look like the original tower. The plains on the plateau of Castile are prone to terrible winter weather, especially snowfall and wind. The snow increased the weight of the tower, thus unbalancing its center of gravity. These conditions caused it to collapse in 1539.

Architect Juan de Vallejo overtook the restoration in the year 1544, but he may have been helped by Francisco de Colonia, the son of Simón who had died in 1542. De Vallejo repaired the tower, coated its columns, and increased the section of the base. This way he decreased its slenderness, that is, the ratio between the height and the supporting surface on the floor. The tower was finished in 1573, but in 1642 a furious hurricane destroyed the eight exterior towers of the cimborio, and seriously damaged the domes around. The repair was also disrupted by a fire in 1644.

Finally, in 1981, another raging storm destroyed some pinnacles and ornaments of the domes. Marcos Rico Santamaría, the architect who restored the cathedral in the 1980s, also reduced the weight of the covering, thus slightly lowering the center of gravity and guaranteeing the stability of the dome. According to this architect, *"the Cimborio is a Cathedral on top of another Cathedral"* [Rico Santamaria 1985].

The vault of the tower is formed by smaller and smaller convex octagons ending in an eight-pointed star polygon. It will be denoted by "8/3", to point out it is a figure formed by straight lines connecting every third point out of the eight regularly spaced vertices of an octagon. The points of another star 8/3 appear on the concave vertices of the polygon. The eight concave vertices of this last star can be joined by segments forming an eight-armed cross (see fig. 4).

Beyond the two main stars on the vault are various triangles, trapezoids and connected squares, forming a grill that covers the roof of the vault. Small circles show trefoils and Greek crosses carved onto them (fig. 5).

Fig. 5. Details of the roof of the vault. Photos by J. Arroyo, reproduced by permission

Mathematical interpretations

A notable proportion in a regular octagon is obtained by dividing its second diagonal and its side (fig. 6), resulting in the $1+\sqrt{2}$ or "silver" proportion or number, here denoted by θ (see [Kappraff 1996] and [Spinadel 1998]). A second proportion is obtained when the length of an octagon's radius is divided by the length of its side (fig. 7), resulting in

$(\sqrt{(2-\sqrt{2})})^{-1}$, denoted by c (see [Redondo Buitrago and Reyes 2008a, 2008b, 2009]). The proportion is called the "*Cordovan proportion*" because it was introduced by Spanish architect Rafael de la Hoz Arderius (see [Hoz 1976]), who investigated the architecture of the Spanish city of Cordoba. θ and c are related by $c = \sqrt{((1+\theta)/2)}$ or $\theta = \sqrt{2}c^2$.

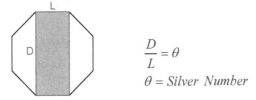

$$\frac{D}{L} = \theta$$
$$\theta = Silver\ Number$$

Fig. 6. Silver rectangle

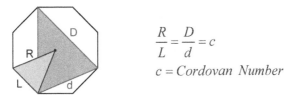

$$\frac{R}{L} = \frac{D}{d} = c$$
$$c = Cordovan\ Number$$

Fig. 7. Cordovan triangles

Note the second (big) rosette is similar to the first (the smaller one), as it results from a central dilation with similarity ratio $c\sqrt{2} = \theta/c$ (fig. 8).

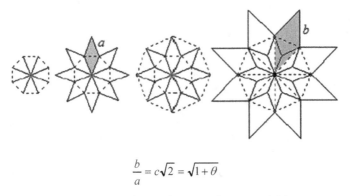

$$\frac{b}{a} = c\sqrt{2} = \sqrt{1+\theta}$$

Fig. 8. Construction of the vault using reflections and dilatations

If a "silver" rectangle whose sides are in a $\theta/1$ proportion, is rotated $45°$ around its centre, an 8/3 star polygon inscribed in the octagon is obtained, as well as an inner octagon (fig. 9).

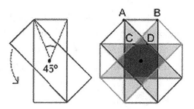

Fig. 9. Rotating a silver rectangle

The ratio AB/CD is the *silver number, θ.* The same star 8/3 follows from a rotation of the two equal sides of a Cordovan triangle (fig. 10).

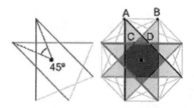

Fig. 10. Rotating a Cordovan Triangle

An 8/2 star is a figure formed by connecting with straight lines every second point of the eight regularly spaced vertices of an octagon. In fig. 11, this 8/2 star is drawn in red, inscribed in a green octagon. The green octagon appears between two blue octagons, whose sides form a geometric sequence with a $θ/c$ ratio.

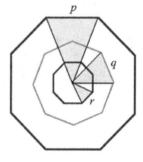

$$\frac{p}{q} = \frac{q}{r} = c\sqrt{2} = θ/c$$

Fig. 11. Star 8/2 in the Cimborio and nested octagons.
Photo at left by J. Arroyo, reproduced by permission

The roof also shows two nested squares inscribed in the quadrilateral formed by the union of a Cordovan triangle and an isosceles right triangle. It illustrates a geometric property:

> Given a quadrilateral PQSR that is the union of a Cordovan triangle PQR, and an isosceles right triangle QRS, the common side QR divides the square ABCD in a √2 rectangle and a $1+θ = 2c^2$ rectangle. The ratio of the areas of the two pieces is equal to θ (fig. 12).

When a square is placed such that three and only three of its vertices intersect a Cordovan triangle *ABC*, another interesting shape is obtained: seven rotations of 45° about its circumcenter O generate a pattern like a grid of stars (fig. 13). It splits in a rosette and a sequence of four nested 8/2 stars (fig. 14).

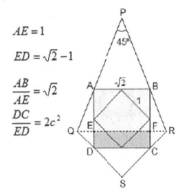

$$AE = 1$$

$$ED = \sqrt{2} - 1$$

$$\frac{AB}{AE} = \sqrt{2}$$

$$\frac{DC}{ED} = 2c^2$$

Fig. 12. Proportions in the quadrilateral details.
Photo at left by J. Arroyo, reproduced by permission

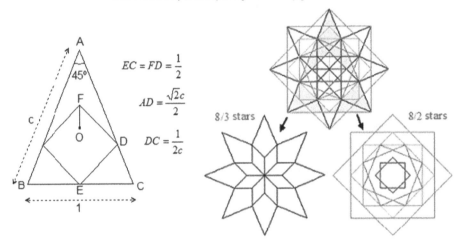

$$EC = FD = \frac{1}{2}$$

$$AD = \frac{\sqrt{2}c}{2}$$

$$DC = \frac{1}{2c}$$

8/3 stars   8/2 stars

Fig. 13. Generator of the starred grid

Fig. 14. The grid of the stars and its subdivision

Placing the orange and blue stars on top of the rosette, the smallest squares of the decorative elements of the stained-glass are locked between them (fig. 15). The two outer stars, in green and orange, define a geometric contraction of ratio √2/2, though the two inner stars, in blue and purple, do not belong to this sequence (fig. 16).

Fig. 15. Decorative elements of the vault   Fig. 16. Contracting the stars

The vertex H of the blue star divides the segment BJ in three segments BH, HI and IJ such that BH=IJ and BJ/HI=θ, since BJ and HI are respectively the second diagonal and the side of the orange octagon with lengths in ratio θ (fig. 17). The purple star plays a similar role when we consider the radius AD and the points B and C. This is a direct consequence of Thales' "intercept" theorem since AF/FG=θ (fig. 18).

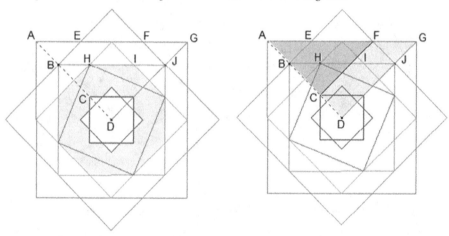

Fig. 17. The blue star and θ Fig. 18. The purple star and θ

The dissection of the Cordovan triangle can be extended by adding more squares (see figs. 19 and 20), allowing to inscribe a rosette inside a tesseract, a tesseract in a rosette, and so on. Fig. 21 shows the fractal character of the pattern.

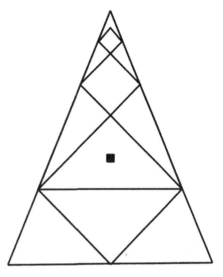

Fig. 19. Trapezoid and triangle on the rosette

Fig. 20. Fractal dissection of the Cordovan triangle

Fig. 21. Nested rosettes and tesseracts

References

Hoz, Rafael de la. 1976. La proporción Cordobesa. *Actas de quinta asamblea de instituciones de Cultura de las Diputaciones.* Córdoba: Ed. Diputación de Córdoba.

KAPPRAFF, Jay. 1996. Musical Proportions at the Basis of Systems of Architectural Proportions both Ancient and Modern. Pp. 115-133 in *Nexus: Architecture and Mathematics*, Kim Williams, ed. Fucecchio (Florence): Edizioni dell'Erba.
http://www.nexusjournal.com/conferences/N1996-Kappraff.html.

REDONDO BUITRAGO, Antonia and Encarnation REYES. 2008a. The Cordovan Proportion: Geometry, Art and Paper folding. *Hyperseeing* May-June 2008: 107:114.
http://www.isama.org/hyperseeing/08/08-c.pdf.

———. 2008b. The Geometry of the Cordovan Polygons. *Visual Mathematics* **10**, 4.
http://www.mi.sanu.ac.rs/vismath/redondo2009/cordovan.pdf

———. 2009. Cordovan Geometrical Patterns and Designs. *Journal of ISIS-Symmetry: Art and Science* 1, 4: 68-71.

RICO SANTAMARÍA, Marcos. 1985. *La Catedral de Burgos. Patrimonio del mundo.* Burgos: Fournier.

SPINADEL, Vera. W. de. 1998. The Metallic Means and Design. Pp. 141-157 in *Nexus II: Architecture and Mathematics*, Kim Williams, ed. Fucecchio (Florence): Edizioni dell'Erba.
http://www.nexusjournal.com/conferences/N1998-Spinadel.html.

About the authors

Dirk Huylebrouck's career in Congo ended after an incident between Belgium and the President of Congo, and so he went teaching in Portugal and later to European Divisions of American Universities, until the first Iraq war drew away his students. He returned to Africa, until a coup d'état during the 1994 genocide epoch ended his contract in Burundi and so he began teaching at the Department for Architecture Sint-Lucas in Brussels and Ghent. Author of the books *Africa + mathematics* about the now famous Ishango rod (in Dutch, translated in French, VUBPress, Brussels, 2004-2008) and *De Codes Van Da Vinci, Bach, Pi & Co* (Academia, 2009), he edits the column "The Mathematical Tourist" in *The Mathematical Intelligencer.*

Antonia Redondo Buitrago has a Ph.D. in applied mathematics from the University of Valencia, Spain. She teaches mathematics in a high school in Albacete (Spain). Her research interests are fractional powers of operators, continued fractions, and the so-called "means" (metallic means, the plastic number and the Cordovan proportion), and their applications in mathematics and education. Her contributions in international journals and congresses are focused on interdisciplinary aspects of mathematics in the domain of the art, design and architecture

Encarnación Reyes Iglesias was born in Torresandino (Burgos). She teaches in the Architecture School of Valladolid, Spain. Her main fields of interest are mathematics in art and architecture: proportions, symmetries, geometry of curves and surfaces, etc. Her most recent books are: *Geometría con el hexágono y el octógono* (Granada, 2009) and *Burbujas de Arte y Matemáticas*, (Madrid, 2009). Her activities in Mathematical Education are focused in updating courses for teachers and mathematical tourist items in several Spanish towns: Salamanca, León, Burgos, etc.

Margarida Tavares da Conceição

University of Coimbra-Centre
for Social Studies
Colégio de S. Jerónimo
3001-401 Coimbra
mmtc@netcabo.pt

Research

Translating Vitruvius and Measuring the Sky: On Pedro Nunes and Architecture

Keywords: Pedro Nunes,
architectural theory, history of
science, architectural treatises,
cosmography, surveying,
Vitruvius, sphere, sundial,
navigation, geodesy, rhumb line,
loxodrome

Abstract. Pedro Nunes is usually associated with the theory of classicist architecture in Portugal, mostly as a result of his declared intention to translate Vitruvius, but over the course of his long career other important theoretical connections between his mathematical procedures and architectural knowledge emerged as well. Although he was famous as a scholar and tutor of princes, his work as a cosmographer shows his indirect contribution to architectural knowledge. More specifically, his teachings on the sphere and use of surveying instruments for measuring spatial coordinates in navigation provide important indications of the link with architecture. These aspects are also demonstrated in his scientific works, such as the *Treatise on the Sphere* and the *Book of Algebra in Geometry and Arithmetic*. Though doubts persist as to his direct relationship with the Vitruvian tradition, there is evidence of knowledge shared between the disciplines, particularly the use of the sundial or nautical needle for orientation and the organization of space.

Scholar and cosmographer

Pedro Nunes (1502-1578) was undoubtedly the most famous Portuguese figure in the history of mathematics, with international influence in his day. He has now become the subject of new research, which has already provided a more solid understanding of his work;[1] indeed, his career and institutional affiliations are important for our appreciation of Portuguese architectural knowledge in the sixteenth century. However, we should perhaps begin by examining his personal background, as mathematics was for him more of a vocation than a profession. In fact, he was responsible for the first usage of the Portuguese word *matemática*, both as a noun ('mathematics') and adjective ('mathematical'), and also, significantly, in the plural (*mathematicas*) in the sense of *artes matemáticas* ('mathematical arts'). This referred to a disciplinary area that was still in formation, a cluster of practical activities united by common bases in geometry and arithmetic, but which did not lead to a profession or to a clearly-defined professional status.[2]

Thus, in labelling Pedro Nunes a mathematician, we are implicitly acknowledging his various professional functions, often exercised simultaneously. His career took place largely within the ambit of the university, but he was also a high-ranking official of the Crown. He may have been born into a Jewish or New Christian family, given the long mathematical tradition amongst Iberian Jews; however, the first documented mention of him occurs with regards to his university training. He studied at the University of Salamanca (1526), but was already in Lisbon when he was appointed royal cosmographer and professor at the Portuguese University, then called '*Estudos Gerais*' (1529).[3] His career really took off when the University was transferred from Lisbon to Coimbra during the reforms launched by King João III in 1537. He was officially appointed professor in 1544, responsible for the chair in mathematics, which had at the same time been established within the course of Medicine, and which now became an autonomous discipline for the first time. He remained responsible for the subject until 1562, the year of his retirement.

As can be seen from this brief summary of his career, he was definitively associated with the university milieu. However, he was also engaged in another kind of high-level teaching, namely the tutoring of princes. In fact, it has been suggested [Leitão 2003: 65] that his professional career may have begun as court mathematician, teaching the king's younger brothers.[4] If so, this would explain his appointment as royal cosmographer and university professor. Indeed, some of what we know about this parallel activity is gleaned from Nunes himself, and is also confirmed by other sources, although in all cases the detail is scanty.

Pedro Nunes's allusions to the scientific aptitudes of some of his disciples shows that reference was already being made to the classical treatises of Antiquity (whether Euclid or Aristotle) in connection with matters related to cosmography, through experimentation with prototypes of navigational instruments. However, it is difficult to generalise on this matter. This is because, given Pedro Nunes's position as royal cosmographer and the physical proximity between the warehouses of India and the royal palace, it is also possible (indeed probable) that there were subsequent contacts with members of the court. This means that it is sometimes difficult to clearly distinguish the position of master from another type of relationship that may have been pedagogical or scientific in nature, such as discussions or informal lessons, which may not be restricted to members of the royal family but extended to young noblemen (*moços fidalgos*).

The likelihood of this has nourished the notion that Pedro Nunes may have taught more extensively at the palace [Moreira 1982, 1998] and that he may even have been tutor to King Sebastião. While this is undocumented institutionally, there is little to contradict it informally, though such lessons would have been open to a very restricted group of people.[5] On the other hand, we should bear in mind that there were other mathematics masters connected to the royal palace who also left behind influential works. These include Domingos Peres,[6] and particularly D. Francisco de Melo,[7] who had an in-depth knowledge of the theoretical sources of architecture.

It is therefore also significant that Pedro Nunes appears to have had a third strand to his teaching activities, for, in addition to his positions as palace tutor and university professor, he also seems to have been engaged in nautical training. In 1529, he was appointed royal cosmographer, becoming chief royal cosmographer in 1547, the first person to occupy this post; indeed, he may have been responsible for drawing up the first regulations concerning it (1559).[8] It appears that particular importance was given to the need for aptitude tests for the manufacturers of nautical instruments, cartographers and seafarers (particularly pilots and captains), and that such tests were necessary for professional accreditation. Thus, the head cosmographer was obliged to give daily mathematics lessons (*lição de mathematica*) to the seamen, though these were apparently not compulsory and were also open to "noblemen" (meaning that they were effectively public lectures). It is not clear how they functioned or for how long, but after 1594 they also became compulsory for architecture apprentices [Conceição 2008: 201-205, 396-401].

Although we do not know if the regulations governing the post were actually applied, it is not difficult to discover the kind of mathematics that was taught in those lessons; the text itself details the materials to be read [Mota 1969: 32-33], and all are related to elementary cosmography and the correct use of navigational instruments. It appears that the lessons involved instruction in the observation of astronomical data through the handling of instruments, and some calculations; that is to say, the emphasis was on practical procedures, such as the use of nautical charts, the astrolabe, cross-staff and quadrant, compass and gnomon. The subjects listed are basic and operational, and do not even include the principles of geometry. Moreover, while the regulations also make provision for the reading of the *Tratado da Sphera* (*Treatise on the Sphere*), this was only for the brighter students, as it was considered to be more advanced knowledge in this context (indeed, it forms the basis of theoretical knowledge in this matter). The information given about exam questions also confirms this focus on practical knowledge and technical procedure; the cumulation of practical experience was much more necessary than theoretical knowledge for the exam, and indeed, experience was the only compulsory condition for admission to it.

While there is no obvious connection with architecture in Pedro Nunes's legacy as scholar and head cosmographer, or indirectly in his professional career, the same cannot be said about the measurement of non-visualizable distances. In fact, it is in the field of surveying (i.e., the measurement of things imperceptible to eye or arm, and the determination of the spatial coordinates of one's location and the place where one is heading) that the measurement of the sphere comes into its own, whatever the scale. And this always involves geometrical procedures; in the case of navigation, the application of the principles of trigonometry. The cosmographer's mathematics lesson was thus concerned with providing a technical apprenticeship in the art of navigation.

We should not, therefore, overlook the possibility of a latent connection between nautical science and architectural practice. The problem, however, is that such a connection (mathematical-geometrical in nature) may only be gauged in the world of practice and materials. If the common denominator is trigonometry, then there is a hiatus on the theoretical level, because the connection is nowhere explained textually. To put it simply, if this connection only existed in practice, that would explain why no direct links with architecture have been found in the career of Pedro Nunes. What appears to exist, though, is a non-explicit connection between nautical science and architecture.

The Tratado da Sphera *and the* Libro de Algebra en Arithmetica y Geometria

Pedro Nunes was essentially a scholar, and it was in this capacity that he produced his theoretical work, most of which is printed. It should be pointed out that this was essentially humanistic scholarship [Tarrió 2002], an aspect that is particularly evident in the philological method that he used to read and translate classical sources. Another feature is his use of various languages; he moves from translations and originals in Portuguese to Latin for international use, and finally, Spanish for a more widespread divulgation of his work.

One of the most interesting aspects revealed by his printed work is the speed with which he apparently gained access to the most recent bibliography of the period. This is related to another important factor, namely his numerous contacts with personalities of international renown (such as Élie Vinet, John Dee, Federico Commandino and Gerardus Mercator, to cite just a few). This obviously operated in reverse, as these circles also acquired knowledge of Nunes's work.[9] In fact, Pedro Nunes is unrivalled by any other Portuguese figure connected to science as regards the quantity of printed works he produced.[10] While this is not the place for an exhaustive list, we may note that his whole oeuvre had something in common – a rigorous and clear use of language, structured around the mathematical discourse and vocabulary of Euclid (i.e., proposition, demonstration and corollary), complemented by the use of drawn figures.

Let us begin by looking at Pedro Nunes's first printed work, the *Tratado da Sphera*[11] [Nunes 2002], an abbreviated title for what was in fact a volume containing five different works. The core of the book is a (partial) translation of three works essential to cosmography. It begins with the most widely-known medieval manual, *Sphaera Mundi* by Johannes de Sacrobosco (John Holywood), which is followed by a translation of the first chapters of *Theoricae novae planetarum* (Nuremberg, 1472, Princeps edition) by Georg Purbach, and the first book of Ptolemy's *Geography*. These three translations were the first to be printed in the Portuguese language.

The first observation to be made is that, rather than a mathematics manual, this is a book of cosmography. It is a collection of texts "for beginners", as Nunes himself acknowledges, suggesting that it had perhaps been compiled at the request of the King; indeed, the three translations are together dedicated to Prince Luís. It was a compendium of basic texts used for teaching at court, although it should be noted that Pedro Nunes did not limit himself to translation and commentary, but also included mathematical demonstrations; nor did he hold back from criticising the classics, while nevertheless acknowledging their essential contributions to the discipline.

The second observation is the most important: this work presents the fundamental theoretical principles for an understanding of the nature of representation of space. As regards Ptolemy's differentiation between geography (the description of the world as a unit) and chorography (the description of particular places), it is clear what differentiates them: the first is quantitative and the second qualitative; one requires measurement and "the institution of mathematics", while the other involves "the painting of places".[12] It is clear that the "painting" or description of places did not interest Pedro Nunes, and that he used it merely as a term of comparison. But this is important to distinguish the nature of the methodological support for each approach: "painting" is here synonymous with representation through drawing of a scene that is in view, perceptible to sight and therefore a kind of sensory knowledge; rigorous geographic representation, on the other hand, requires geometric demonstration and the use of the appropriate instruments. But geographical surveying, and the kind of knowledge of the earth acquired through observation of the heavens, require what we today call geographic coordinates, i.e. a means of knowing where a point or place is located without directly observing it, in order for it to be represented in another operation.

This leads directly to the heart of the problem of representation of space, cartographic demarcation and the mathematical abstraction that this implies. The title of Chapter XXI, "On the things that need to be retained when describing the orb on a plane surface",[13] makes clear the nature of the cartographer/cosmographer's problem. It is essentially the same problem faced by architects and/or topographic engineers, namely understanding and representing spaces, buildings and machines already existing and (above all) to be planned. The process of architectural and engineering design requires abstract codified graphic representation and the ability to predict, plan and calculate (in all senses of the word) a structure as a result of and in response to a problem, need or force.

This matter of cartographic representation is continued in two small treatises by Nunes himself, which follow the three translations: *Treatise... on some Navigational Doubts* and *Treatise... Defending the Sea Chart: with the regulation of height* (in [Nunes 2002]).[14] These constitute his original innovative contribution to the body of scientific writings on navigation and include important information about the problem of representing spherical bodies on a plane surface. Each of these two texts (but particularly the second) is part of a quest to find the most accurate method of drawing maps to aid navigation.

His controversies with the seamen, more cited than known in detail, are reflected here. In several passages, Pedro Nunes criticises the pilots' scanty knowledge and lack of cartographic skill. The crux of the matter seems to have been their intellectual limitations and inability to process abstract thought. The mathematician's discursive clarity is evident both textually and graphically, with geometric demonstrations of the rhumb line and the reason for preferring great circles in navigation. In fact, it is here that he expounds his "discovery" of the rhumb line (later known as the loxodrome): that which "is not a circle nor a straight line" but "rather a curved irregular line" (fig. 1), explaining the reasons why, in practice, "sailors rarely get it right".[15]

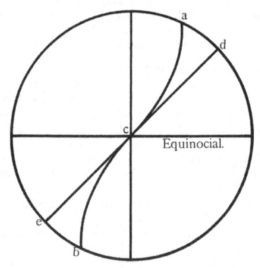

Fig. 1. Graphic representation of the so-called "rhumb line" or "loxodrome", from the *Tratado da Sphera* [Nunes 2002: 113]

Despite his criticism of sailors, Nunes was nevertheless very aware of the radical innovations involved in the Portuguese discoveries, "which were not achieved by trial and error, but by our highly-trained seamen equipped with instruments and rules of astrology and geometry".[16] In fact, it is clear from his discourse that his overriding concern was with the rigour of that knowledge, for successful navigation depended upon it.

Basic learning is again the subject of his last printed book,[17] the *Libro de algebra en arithmetica y geometria...* [Nunes 1950], which is the most interesting for architecture and fortification. It is presented as the last in a series of publications of original works by Pedro Nunes, and takes up an important aspect of his first book of 1537, namely education. The use of the Spanish language indicates that this was a work of divulgation, clearly aimed at a much more extensive public than the university scholar. It is effectively a book for self-study, starting with simple matters and increasing in complexity, and is not restricted to a compilation of basic rules; indeed, it goes into the matter in some depth, by applying the principle of algebra to arithmetic and geometry.

The introductory pages explain that the book is based upon a text initially written in Portuguese some thirty years before, which circulated in manuscript form. Nunes was now obliged to revise and extend the work, updating it to take account of more recent publications, such as those by Girolamo Cardano (1501-1576) and Niccolò Fontana, known as Tartaglia (c. 1500-1557), who were embroiled in a polemic. In fact, his criticisms were mainly directed at Tartaglia's *Quesiti et inventioni diverse* and Cardano's *Practica arithmeticae et mensurandi singularis* (Milan, 1539) and *Ars Magna* (*Artis Magnae, sive de regulis algebraicis...*, Nuremberg, 1545), although Luca Pacioli (1445-1517) also comes under fire.

Tartaglia undertook the first translation of Euclid's *Elements* (Venice, 1545) and also of Archimedes' work (Venice, 1543), but his work was especially relevant to military architecture. He had already published a work on fortification, *La nuova scientia* (1537, re-edited in 1550), a text in the form of a dialogue, which mixes allegory and mathematics, but which nevertheless constituted the first systematic approach to ballistics as a basis for the planning of fortifications. It was his book *Quesiti et inventioni diverse*, first published in Venice in 1546, and re-edited with additions in 1554, that was the most cited and criticised by Pedro Nunes, along with the *General trattato di numeri et misure* [1556]. The fact that Tartaglia's work was known and studied in Portugal is naturally very important for the theory (and possibly also for the practice) of fortification in the Portuguese context. However, Pedro Nunes's references to it are strictly mathematical, and it is not possible to speculate about the extent to which Tartaglia's work was studied in Portugal or its implications for military engineering prior to the seventeenth century.[18]

Both Joaquim de Carvalho and Henrique Leitão, in various works, mention that the *Book of Algebra...* circulated widely, partly due to its influence on the Jesuit, Christoph Clavius. In this work, Nunes demonstrates that he belongs to the restricted number of early mathematicians who truly made use of the principle of algebra, that is, using the notion of the unknown variable (*cosa ignota*) to find a solution to problems that could not be solved through arithmetic or geometry (see "Notas e comentários" to the *Libro de Algebra* by Victor Hugo Duarte de Lemos in [Nunes 1950: 469-498]). It should also be pointed out that, within the Europe-wide chain of transmission of scientific knowledge, Nunes's algebra was in turn analysed and criticised by the Flemish mathematician and military engineer, Simon Stevin (another important name in the development of the military treatise, and pioneer of the Dutch School, whose influence seems to have arrived in Portugal in the first half of the seventeenth century).[19]

It is important to highlight this less obvious aspect of the reception and contraposition of ideas, which used the comparison of data as an intrinsic and indispensable part of scientific progress. Indeed, such criticism seems to be part and parcel of the mathematical mentality (demonstration), while scientific proof is intrinsically mathematical in nature, implying, in general terms, a method. When Pedro Nunes granted a "scientific place" to Architecture,[20] he was awarding it membership status in accordance with the application of a method: that of the mathematical sciences.

It should also be pointed out that, alongside Euclid, the work of Archimedes was also fundamental for physics and mechanics. In this sense, we understand how the structure of the treatise, which is divided into three main parts, makes it accessible to non-specialists, presenting information in an organized methodical sequence. The basic definitions are given first, with the statement of rules and respective demonstrations, while the second part deals with the application of those rules to arithmetical operations, roots and "proportions". The third part deals with equations applied to cases of arithmetic, and (more importantly for engineering) to cases of geometry, presenting sequentially the square, rectangles, triangles, rhombuses, trapeziums, pentagons "and other many-sided shapes". If we compare this with Book VI of Tartaglia's *Quesiti e inventioni diverse*, we can see the advance that had been made with the possibility of solving cubic equations (by definition, equations that resolve three-dimensional spatial problems, i.e., geometric).

In addition to the effectiveness of the operations involved in the calculation and planning of defensive structures, intellectual training in mathematics would itself influence the disciplinary contours of fortification, in that similar methods were used, whatever the scale of the object under study (including the city and the territory). In general terms, this was Pedro Nunes's great indirect contribution to the science/art of military architecture and indeed to architecture as a whole. It was a literally magisterial contribution, made in his capacity as teacher (*magister*). The very concept of science as knowledge supported by and derived from mathematical demonstration – a consideration that connects all his writing and is often made explicit in his discourse [Leitão 2002b: 56-58] – soon became an integral part of architectural training, particularly for military engineering and fortification. This meant that there was now a theoretical component of some kind.

Translating Vitruvius and other attempts

Pedro Nunes's theoretical output was not limited to his printed works, but also extended to various manuscripts (at least one extant, some attributed and others now disappeared).[21] One of those lost is obviously of fundamental importance for the theory of architecture: the translation of Vitruvius. Indeed, it is as interpreter of Vitruvius that Pedro Nunes has occasionally appeared in connection with architecture, though this is ultimately a dead-end, as it comes up against something that does not exist. However, it is worthwhile summarising the information that we have on the subject. We know first-hand from Nunes himself that there was an intention and an action; in the dedication of *De crepusculis* [Nunes 2003], dated October 1541, Nunes mentions that he had worked on the translation and/or commentary of Vitruvius's treatise *De architectura* at the request of King João III.[22]

The importance of this in the cultural context of the 1540s is only demonstrable by comparing it with other attempts to translate essential texts for the theory and history of architecture. The interest in books on architecture may be gauged by the re-publication of *Medidas del Romano* by Diego de Sagredo (ca. 1490-ca. 1528), which was printed in 1526 in Toledo and ran to three editions in Lisbon, between 1541 and 1542.[23] A little later, André de Resende, a humanist close to Erasmus, was ordered by the king to translate a book of architecture, probably Alberti's *De re aedificatoria* in around 1551 or 1552.[24] Prior to this, André de Resende had translated (ca. 1542-1543), "two books about aqueducts", possibly the *De aquaeductibus urbis romae* by Sexto Júlio Frontino, one of the appendices that accompanied editions of Fra Giocondo [Moreira 1983: 42; Moreira 1991: 351-352; Moreira 1995a: 346; Deswarte, 1992: 171]. There is also indirect evidence of another translation, the book about the fortification of cities by

Albrecht Dürer [1527], which may have been translated into Portuguese in 1552 by Isidoro de Almeida [Moreira 1983: 343], a soldier with some vocation for architectural design.

If Pedro Nunes's translation of Vitruvius actually existed (in 1541-1542) and was printed, it would have been the first vernacular version outside Italy[25] [Moreira 1998: 385]. But was it ever finished? Joaquim de Carvalho, noting the meaning of specific terms used by Pedro Nunes, suggests that a translation with critical explanations did exist (the word *interpretatio* recurs), though he doubts that the task was concluded (see [Nunes 2003: 315-317]). However, Henrique Leitão [2002 b: 45] points out that Nunes mentions Vitruvius on certain occasions, though always within the context of the scientific themes covered in Book IX of *De architectura*, and argues that there is sufficient evidence to assert that the text existed and, like other lost unpublished works, was known outside Portugal [Leitão 2002: 65-66; 2002a: 122]. That is to say, in addition to Pedro Nunes's own words, there is also some evidence elsewhere, ill-defined but nevertheless visible.

The first piece of evidence is more of an indirect proof than an indication. The inventory of the books belonging to the Spanish architect Francisco de Mora mentions "a Lusitanian, Pedro Nuñez, of architecture and navigation",[26] which would certainly be a reference to Nunes's translation of Vitruvius and to some text about nautical science. A second clue, also cited by Henrique Leitão [2002a: 122] after Rafael Moreira [1995b: 51], is the fact that the humanist Walther Hermann Ryff (ca. 1500-ca. 1548), who undertook the first German translation of *De architectura libri decem* (known as the *Vitruvius Teustsh*, Nuremberg, 1548), mentions the name of Pedro Nunes as one of the authorities on the matter.[27] Ryff was both a physician and a mathematician, which would combine well with the study of the work of Pedro Nunes or contact with some of his works.

There is yet another clue, uncertain though very interesting. Henrique Leitão mentions an unresolved question concerning Francesco Maurolico (1494-1575), a Sicilian mathematician who, in his *Cosmographia* (1543) on the subject of wind names and directions, cites an opinion that he attributes to Pedro Nunes [Leitão 2002a: 117]. Given the date of that printed reference and the expression used by Nunes with respect to *interpretationem vitruuii*, it is likely to refer to the winds from Book I of *De architectura*, which the Portuguese mathematician was clearly familiar with. Vitruvius's first book is undoubtedly the most far-reaching in cultural and disciplinary terms, and contains the basic definitions; it could not have been unknown to the humanist Pedro Nunes. Moreover, Chapter 6 of that book is entirely dedicated to town planning, and street layout had to take account of wind types, particularly the direction of the eight main winds. This is, in fact, one of Vitruvius's most cited topics in architecture theory due to the matricial induction of the geometric octagon.

On the other hand, even if a translation (partial or otherwise) had existed and circulated, Vitruvius appears to have made no impact whatsoever on the extensive writings of Pedro Nunes. Direct or indirect quotations from him are very scarce, and when they occur, it is only in connection with cosmographic matters.

Moreover, in the chapter of *Tratado da Sphera* concerning the measurements of the sphere [Nunes 2002:14], Pedro Nunes actively suppresses Vitruvius amongst the authorities cited in Sacrobosco's text, referring only to Ambrosius Theodosius Macrobius and Eratosthenes. However, this omission also occurs in some other versions of this medieval treatise.[28] The context of that annotation is the chapter about "the distribution of squares and streets" (i.e., urban layout) and the sentence appears after an explanation of the use of the gnomon and the different directions of streets and winds. This may lead to two interlinked hypotheses: firstly, Pedro Nunes may have avoided mentioning Vitruvius because he considered him irrelevant, on the grounds that he drew his information from Erastosthenes. This would in turn imply that Nunes had read and studied Vitruvius directly and realised that his opinion was not his own but that he was referring to another source.

But, in this volume of *Tratado da Sphera*, there is another explicit reference to Vitruvius in another context. It comes at the beginning of the *Treatise...on Some Navigational Doubts* concerning the explanation of rhumb lines and is worth transcribing:

> Because it is clear that: drawing a meridian line on the horizon: by the art that Vitruvius brings to this: that line is called the north-south rhumb line: and crossing this line with another perpendicular to it: we will have the east-west rhumb line: and thus the whole circle of the horizon will be divided into quarters, each of ninety degrees. And this is thus represented by the needle with which we navigate: and also by any of the needles that are painted on the map.[29]

This sentence by Pedro Nunes is revealing. Rafael Moreira [1995b: 51] believes that the reference to the "art of Vitruvius" alludes to the *analemmma*, or structural lines of the horizon, which concerns Chapter 1 of Book IX of *De Architectura,* the book dealing with "knowledge of astronomy and gnomonics". But this suggestion does not annul the knowledge that Pedro Nunes demonstrates of Chapter 6 of Book I, mentioned above, and the sentence itself reveals his understanding of the Vitruvian principles expressed there.

Pedro Nunes is clearly speaking of cosmography in general terms. This means that the sense is by definition spatial, so much so, in fact, that when he refers to the navigation needle, he compares it to other "needles painted on the map" (the windrose), but which could also be "needles" on the ground, such as the gnomon or sundial. In this specific passage citing the *art* of Vitruvius (and Pedro Nunes does not usually make gratuitous assertions), what is at stake is the general principle of organization of space and the instrumentation used for orientation. This coincides with the principle of territorial division and demarcation that occurs during the founding of a city, a subject that completely occupies Book I, alongside the definition of architecture and its elements.

Moreover, what Pedro Nunes effectively describes is the basic scheme of geometry, the relationship of the perpendicular to the cardinal geographic points. This cross scheme constitutes a kind of common heritage shared, on different scales, by geography and astronomy, and the idea of the city.[30] This observation is related to Pedro Nunes on a specific point: the instrumental needs of nautical orientation and, in that sense, the reading that the mathematician could not have avoided making from Book I of Vitruvius's architecture treatise, which itself constitutes a repository of classical knowledge on this matter up to the first century. The notion that Pedro Nunes was only interested in Vitruvius's more technical books may confuse or mask the issue. Given the sixteenth-century obsession with architecture and its personification in Vitruvius, it is almost impossible that Pedro Nunes had not read it thoroughly. And from it he retained what interested him, as is natural.

But, even if Pedro Nunes had only retained the scientific aspects of Vitruvius, such as information about gnomonics, this is not clearly reflected in his own texts.[31] And we should not forget how crucial was the construction of reliable clocks, as Pedro Nunes understood, since a rigorous method of measuring time was required to determine longitude.[32] This note is important for the verification of the very meaning and use of gnomonics, transforming the gnomon into the common principle underlying all orientation in time and space (by means of the sun's course, a criterion invested with essentiality). This served not only as the basis of the clock (sundial), but also for marking the geometric centre or navel of the city, which regulated the division of territory and urban layout. Indeed, this principle could be applied whatever kind of habitat was involved.

The measurability of space: the sphere and architecture

The medieval notion of a closed world, where space was a symbolic given, gradually gave way to the concept of an infinite universe, in which space could be measured in an increasingly accurate way. However, this change was only really consolidated in the eighteenth century. Underlying it was the development of new conceptual tools that gave control of the spatial dimension and the abstract representation of reality, concrete or virtual. These were mostly developed in the mathematical disciplines, of which geometry was the touchstone. Thus, the discovery of the sphere of the earth as a space of quantification and measurability provided the context for scientific (or more precisely, pre-scientific) knowledge in the era of the Discoveries, proving decisive for the "opening-up" of the world. It also enabled representation from the point of view of the city and architecture.

It is not by chance that there are unequivocal historical references in the first four decades of the sixteenth century to the appearance of architectural designs, in which it is even possible to identify different genres of representation, such as drawings and urban maps. It may also not be chance that there is not the slightest trace of their material existence in Portuguese archives, throwing doubt upon their use and effective conceptual utility [Rossa 2002: 380-381]. Nor is it by chance that sixteenth-century European culture saw a wave of treatises on surveying and instrument-making. There were some reverberations in Portugal, although these have only really been documented and studied in connection with nautical science (this aspect will not be developed in any detail here). The way that some astronomical instruments could be used for both navigational and geodesic purposes is not unknown to historians of science.

However, this deep-rooted relationship between architecture, town planning and nautical science seems to have taken place on another, less explicit level. Recent data from the research of João Horta [2006] uses mathematical analysis to demonstrate this connection in a way that only becomes operational on a case-by-case, city-by-city or object-by-object basis. In terms of scale, it is as if, from the horizon of the sphere, a microscope were used to see the ground. This is a fundamental question, as it means that mathematical language may be used to reconstruct the planning process using a process of geometrical analytic deconstruction – an aspect of research where traditional historical analysis of theoretical texts on town-planning proves insufficient or inconclusive.

In short, in Horta's work [2006: 79], a direct link is made between the three-dimensional logic of nautical geometry (which is necessarily cosmographic or Sphere-based) and the plane surface geometry of architecture and town planning. The abacus, diagram or grid, used as a tool for town planning and architecture (including naval architecture), shares a common matrix with the so-called compass rose or wind rose. In other words, an archetypal octagonal generative system was used as a basis for planning because it contains the principle for determining geometric solutions/relations.

Thus, it may be seen how this generative schema coincides with the so-called Vitruvian octagon, which is also deduced from the cardinal points. This clearly shows how classical archetypes and proportional measurement systems may be used to generate variable geometric patterns that are coherent in their common matrix. Also significant is the fact that the same matrix is evoked by Pedro Nunes for the cartographic design of the terrestrial sphere, as mentioned above [Nunes 2002: 106], thereby recovering Vitruvius's legacy, itself an archetypal treatise.

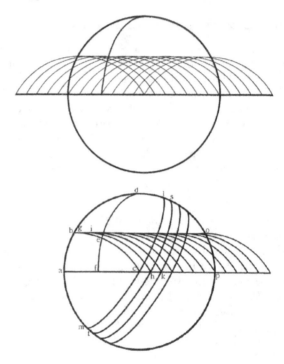

Fig. 2. Cosmographic diagrams based on trigonometry, from the *Tratado da Sphera* [Nunes 2002: 108-109]

To be precise, the instrumental capacity of the windrose diagram is based upon trigonometric principles inherent in the spherical nature of the earthly globe. It effectively constitutes a trigonometric grid that can be adapted to town planning in specific cases (fig. 2). That is to say, the same trigonometric principle, calculated diagrammatically, using an abacus or with a combination of both, is always based upon a double quadrature or octagonal base, whose lines of angular metric coordination may be used on the geographic, urbanistic or architectural scale. This means that it could virtually generate various proportional systems of differing degrees of complexity.

To sum up, João Horta has demonstrated that it is possible to use nautical charts or a simple abacus scheme to design on paper or on the ground in a geometrically coherent way. It was probably not architects or engineers, but rather pilots, foremen and soldiers, people with little sophisticated literary or theoretical knowledge but with elementary mathematics, who were responsible for navigation, docking and building. More important than the architecture and fortification treatises that circulated during the sixteenth century would have been "the use of calculation tools, like the set square, ruler, compass or design manual, instruments whose worth was already proven and valued, and which circulated unpretentiously. But this labour is never given credit. Perhaps that is why there seems to have been no documental evidence of its use" [Horta 2006: 125].

This almost undocumented effort will also have been related to the need for a representation of the Sphere and with the conception of the universe itself. The development of pre-Cartesian philosophy, still firmly anchored in Aristotelianism, though gradually advancing towards geometric rationality, frames architectural and town planning theory within a broader intellectual process. The principles underlying architecture and town planning, whether expounded in treatises or not, are also related to the mathematization of reality. The ability to measure space and the things in it, describe it textually and represent it cartographically came about not only through experimentation (as a way of validating knowledge) and advances in nautical science, but also, crucially, through the concomitant development of the mathematical sciences, which laid the foundations for the rest.

Translated from the Portuguese by Karen Bennett

Notes

1. The bibliography on Pedro Nunes is very extensive (see [Leitão 2002] for a recent overview). An unfinished project by the Lisbon Academy of Sciences to publish Pedro Nunes's complete works (begun in 1940 and edited by Joaquim de Carvalho) has recently been resumed with the support of the Calouste Gulbenkian Foundation, with Henrique Leitão as general editor. Since 2002, Volumes I, II, III, IV and VI out of a total of eight planned have been published.
2. This was shown clearly in a recent exhibition and catalogue, *Compass and Rule: Architecture as Mathematical Practice in England* [Gerbino and Johnston 2009].
3. He was later awarded a Doctorate in Medicine (1532) by this institution, although it is not known if he ever actually became a practising "physician".
4. He certainly taught Prince Luís (1506-1555), Prince Henrique (1512-1580) and Prince Duarte (1515-1540), the last of whom had significant connections with the Ducal court of the Braganças, based in Vila Viçosa. For further information on this, see [Conceição 2008: 196-197].
5. One of those characters was undoubtedly D. João de Castro (1500-1548), Viceroy of India, a multi-faceted figure, whose works reveal an awareness of the importance of Portuguese geographical and cosmographic discoveries in relation to those of the ancients, which he systematically criticises. The same attitude is evident in his familiarity with Vitruvius's *De architectura* and the comparison he makes with Indian architecture, also considered to be "ancient". For a more in-depth study of this issue, see [Deswarte 1992: 37-54] and [Moreira 1995b].
6. He left a handwritten version of part of Euclid's *Elements*, a fragmentary simplified translation meant to be used for elementary teaching purposes. The manuscript was published by Luís de Albuquerque [1968: 121-198].
7. This more complex figure (1490-1536) left original commentaries on texts by Euclid and Archimedes, dating from between 1514 and 1517 [Leitão: 2005: 296; Albuquerque, 1968: 2]. In his inaugural lecture delivered in 1535, he reveals the influence of architecture theorists, citing Vitruvius, though basing himself on an interpretation resulting from Alberti's *De re aedificatoria* (see *Oração de Sapiência* (*Oratio pro rostris*), Miguel Pinto de Meneses, trans., A. Moreira de Sá, ed., Lisbon: Centro de Estudos de História da Filosofia,1956, pp. 154-157).
8. Here too the extent of his involvement is not known in any detail, and most of our knowledge has been acquired indirectly, via an allusion made to it in the "Regulations concerning the post of Head Cosmographer" (*Regimento do Cosmógrafo-mor*) of 1592. This was published and

studied by Avelino Teixeira da Mota [1969] and has been taken upon again more recently by Rita Cortez de Matos [1999].

9. A well-documented "dissemination network" existed with regard to the works of Pedro Nunes even while he was alive. It was particularly active in the first decades of the next century, largely due to the Jesuit school of mathematics (see the work of Henrique Leitão, particularly [2002a]).

10. His printed work included *Tratado da Sphera* (Lisbon, 1537), *Astronomici Introductori de Sphaera Epitome*, (before 1541); *De crepusculis...* (Lisbon,1542); *De erratis Orontii Finaei...* (Coimbra, 1546; *Petri Nonii Salaciensis Opera...* (Basel, 1566); *Libro de Algebra en Arithmetica y Geometria...* (Antwerp, 1567). Some of the volumes contain different texts, including translations, originals and re-editions, sometimes dedicated to different people.

11. Published in Lisbon in 1537, the frontispiece read: *Tratado da Sphera com a Theorica do Sol e da Lua. E ho primeiro livro da Geographia de Claudio Ptolomeo Alexãdrino tirados nouamente de latim em lingoagem pello Doutor Pero Nunez cosmographo del Rey dõ Ioão he terceyro deste nome nosso Senhor. E acrece[n]tados de muitas annotações e figuras per mays facilmente se podem entender. // Item dous tratados que o mesmo Doutor fez sobre a Carta de marear: Em os quaes se decrarão todas as principaes duuidas da nauegação. Cõ as tauoas do mouimento do sol: e sua declinação. E o Regimento da altura assi ao meyo dia: como nos outros tempos.*

12. *... pintura dos lugares: e nenhum homem sera Corographo: se não for pintor ...* [Nunes 2002: 64].

13. *Das cousas que se ham de guardar na descripção do orbe em plano* [Nunes 2002: 89-90].

14. These two treatises were *Tratado... sobre certas duuidas da nauegação* and *Tratado... em defensam das cartas de marear: com o regimento da altura...* , reprinted in 1566, in a Latin version *De duobus problematis circa nauigandi artem liber unus* and later included in the book *De arte atque ratione nauigandi libri duo*, 1573.

15. *...qual não he circulo: nem linha direita ... antes he hua linha curua; e yrregular ..., que os nauegantes poucas uezes acertam* [Nunes 2002: 112].

16. *... que nam se fezeram indo a acertar: mas partiam os nossos mareantes muy ensinados e prouidos de estormentos e regras de astrologia e geometria: que sam as cousas de que os Cosmographos ham dãdar apercebidos* [Nunes 2002: 121].

17. *Libro de Algebra en Arithmetica y Geometria...*, Antwerp, printed in *"casa de los herderos d' Arnoldo Birckman a la Gallina gorda"* or *"em casa de la Biuda y herederos de Iuan Stelsio"*. The fact that two printers were responsible for different copies of the same edition was a very unusual event, which has not been satisfactorily explained. One of the copies preserved in the National Library of Portugal (BNP, Res. 734 P.) belonged to Luís Serrão Pimentel (1612-1679), head cosmographer and military engineer, who also wrote several treatises.

18. Pedro Nunes was probably not the only one reading Tartaglia's books. In any case, the *Quesiti et inventioni diverse* is largely concerned with fortification and military art: artillery (Books I-III); tactics (Book IV); geodetics or land surveying (Book V) and urban fortification (Books V-VII). Of the nine books, only the last three are devoted to purely mathematical subjects.

19. Simon Stevin, *L'Arithmetique... contenant les computations des nombres Arithmetiques ou vulgaires; Aussi l'algebre, auec les equations de cinc quantitez...* (Leiden, 1585, 1625 and 1634), quoted by Joaquim de Carvalho in his "Anotações histórico-bibliográficas" to the *Libro de Algebra...* in [Nunes 1950: 413-467].

20. *De todollos Liuros que nas Sciencias Mathematicas tenho composto, muito alto & excellente Principe, nenhum he de tanto proueito como este de Algebra, que he conta facil & breue para conhecer a quantidade ignota, em qualquer proposito de Arithmetica & Geometria, & em toda a arte que usa de conta & de medida, como sam Cosmographia, Astrologia, Architectura & Mercantil. E posto que os principios desta subtilissima arte sejam tirados dos Liuros elementarios de Euclides, nam se pode porem sem ella ter a practica dos mesmos liuros, & dos de Archimedes* [Nunes 1950: XIII].

21. The only surviving manuscript is *Defensão do Tratado da Rumação do Globo para a Arte de Navegar*, published by Joaquim de Carvalho [Nunes 1952].

22. *... sed ut occasionem aliquam nanciscerer excusandi me quod interpretationem Vitruuii tamdium sim moratus: nam prae aduersa ualetudine inchoatum opus et supra quam dimidiatum non absolui* [Nunes 2003: 6].

23. The first Lisbon edition was dated June 1541, the second January 1542 and the third June 1542. All were printed by Luís Rodrigues [Moreira 1995a: 344], who also published Pedro Nunes.

24. *Livro de Arquitectura, tradução da Arquitectura de Leão Baptista.* Rafael Moreira suggests that this translation of 1550-1551 had been done from Cosimo Bartoli's Italian version (Florence, 1550 [Moreira 1983: 342]. See [Deswarte 1981: 238; 1988: 333; 1992: 175]; see also [Conceição 2008: 129].

25. Of Vitruvius, the first printed Italian translation was by Cesare Cesariano, 1521; the first printed French translation, by Jean Martin, 1547; the first German translation, by Walther Ryff, 1548; the first printed Spanish translation, by Miguel de Urrea, 1582; see "timeline" in [Conceição 2008].

26. *... um Pedro nuñez lusitano de arquitectura y de nabegacion ...* [Leitão 2002a: 122].

27. Prior to the German translation, Ryff had already published the Latin version of *De architectura*, in 1543.

28. Joaquim de Carvalho in [Nunes 2002: 238]. Rafael Moreira [1995b: 51] also refers to this omission and notes that Pedro Nunes may even have highlighted this information source, which is part of Book I, Chapter 6 of *De architectura*.

29. *Porque cousa clara he: que fazendo no orizonte huma linha meridiana: pella arte que Vetruuio pera isso traz: a tal linha se chama o rumo de norte sul: e atravessando linha com outra perpendicular: teremos o rumo de leste oeste: e assi ficara repartido todo o circulo do orizonte em quartas cada huma de nouenta graos. E isto se representa assi pella agulha com que nauegamos: como tambem per qualquer das agulhas que na carta se pintam* [Nunes 2002: 106].

30. This is best demonstrated by *The Idea of a Town* [Rykwert 1988], which identifies a ritualistic-symbolic connection between the form of the city and the religious-transcendant conception of the world. This is encapsulated in the symbol of the cross inside a circle — ⊕ — which could be both a cosmological sign, and (in certain pre-Classical and Classical civilizations) an actual ideogram for the word/concept of the city.

31. In *De erratis orontii finaei...*, while opposing the Frenchman's proposals for the construction of the sundial (see Henrique Leitão, "Anotações..." in [Nunes 2005: 318 and 394]), Nunes explains the basic principles of gnomonics in relation to horizontal and vertical sundials. However, there is no reference here to Vitruvius, which suggests he did not give much importance to the cited Book IX.

32. For this reason, the study and use of the sundial is listed in the *Regulations concerning the post of Head Cosmographer* as a subject to be taught to pilots [Mota 1969].

References

ALBUQUERQUE, Luís de. 1969. *Fragmentos de Euclides numa versão portuguesa do séc. XVI.* Lisbon: Junta de Investigações do Ultramar.

CONCEIÇÃO, Margarida Tavares da. 2000. A Praça de Guerra, aprendizagens entre a Aula do Paço e a Aula de Fortificação. *Oceanos* **41**: 24-38 (Lisbon: Comissão Nacional para as Comemorações dos Descobrimentos Portugueses).

———. 2008. Da cidade e fortificação em textos portugueses (1540-1640). Ph.D. dissertation (Theory and History of Architecture), Faculty of Sciences and Technology, University of Coimbra.

DESWARTE, Sylvie. 1988. Francisco de Holanda ou le Diable vêtu à l'italienne. Pp. 327-345 in *Les Traités d'Architecture de La Renaissance*, Jean Guillaume, ed. Tours: Picard.

———. 1988. La Rome de D. Miguel da Silva (1515-1525). Pp. 177-307 in *O Humanisno Português, 1500-1600* (Primeiro Simpósio Nacional, Lisbon, 1985). Lisbon: Academia de Ciências de Lisboa.

————. 1992. *Imagens e Ideias na Época dos Descobrimentos, Francisco de Holanda e a Teoria da Arte*. Lisbon: Difel.

DÜRER, Albrecht. 1527. *Etliche Underricht zu Befestigung der Stett, Schloz und Flecken*. Nuremberg. (Latin edition, *De urbibus, arcibus, castellisque condendis, ac muniendis rationes...*, Joachim Camerarius, trans. Paris: Christian Wechel, 1535.)

GERBINO, Anthony and Stephen JOHNSTON. 2009. *Compass & Rule: Architecture as Mathematical Practice in England, 1500-1750*. New Haven – London: Yale University Press - Museum of the History of Science Oxford.

HORTA, João Manuel Gomes. 2006. Vila Real de Santo António, forma limite no urbanismo histórico português. Ph.D. dissertation, Faculty of Humanities and Social Sciences, University of Algarve.

LEITÃO, Henrique, ed. 2002. *Pedro Nunes, 1502-1578. Novas terras, novos mares e o que mays he: novo ceo e novas estrellas*. Exhibition catalogue. Lisbon: Biblioteca Nacional.

————. 2002a. Sobre a difusão europeia da obra de Pedro Nunes. *Oceanos* **49**: 110-128 (Lisbon: Comissão Nacional para as Comemorações dos Descobrimentos Portugueses).

————. 2002b. Pedro Nunes, leitor de livros antigos e modernos. Pp. 31-58 in *Pedro Nunes e Damião de Góis – Dois Rostos do Humanismo Português, Actas do Colóquio*, Aires Augusto do Nascimento, ed. Lisbon: Guimarães Editores.

————. 2003. Para uma biografia de Pedro Nunes: o surgimento de um matemático, 1502-1542. *Cadernos de Estudos Sefarditas* **3**: 45-82 (Lisbon: Universidade de Lisboa).

————, ed. 2004. *O Livro Científico dos Séculos XV e XVI. Ciências Físico-Matemáticas na Biblioteca Nacional*. Lisbon: Biblioteca Nacional.

MATOS, Rita Cortês de. 1999. O cosmógrafo-mor: o ensino náutico em Portugal nos séculos XVI e XVII. *Oceanos* **38**: 55-64 (Lisbon: Comissão Nacional para as Comemorações dos Descobrimentos Portugueses).

MOREIRA, Rafael. 1998. Um Tratado Português de Arquitectura do Século XVI (1576 - 1579). *Colectânea de estudos Universo Urbanístico Português 1415-1822*. Lisbon: Comissão Nacional para as Comemorações dos Descobrimentos Portugueses, pp. 353-398.

————. 1983. Arquitectura. Pp. 305-352 in Catalogue of the XVII Exposição de Arte, Ciência e Cultura, Arte Antiga –I. Lisbon.

————. 1991. A Arquitectura do Renascimento no Sul de Portugal, a Encomenda Régia entre o Moderno e o Romano. Ph.D. dissertation, Faculty of Humanities and Social Sciences, Universidade Nova de Lisboa.

————. 1995a. Arquitectura: Renascimento e Classicismo. *História da Arte Portuguesa*, Paulo Pereira, ed. Lisbon: Círculo de Leitores, vol. II, pp. 303-375.

————. 1995b. D. João de Castro e Vitrúvio. *As tapeçarias de D. João de Castro*. Exhibition catalogue. Lisbon: Comissão Nacional para as Comemorações dos Descobrimentos Portugueses – IPM, pp. 51-56.

MOTA, Avelino Teixeira da. 1969. *Os regimentos do cosmógrafo-mor de 1559 e 1592 e as origens do ensino náutico em Portugal*. Lisbon: Junta de Investigações do Ultramar.

NUNES, Pedro. 1950. *Libro de Algebra en Arithmetica y Geometria* (1567). *Obras*, vol. VI, Joaquim de Carvalho, et al, eds. Lisbon: Academia das Ciências de Lisboa, Imprensa Nacional de Lisboa.

————. 1952. *Defensão do tratado da rumação do globo para a arte de navegar*. Joaquim de Carvalho, ed. Coimbra: *Revista da Universidade de Coimbra* **17**.

————. 2002. *Tratado da Sphera Astronomici Introductorii de Sphaera Epitome* (1537). *Obras*, vol. I, Henrique Leitão, et al., eds. Lisbon: Academia das Ciências de Lisboa, Fundação Calouste Gulbenkian.

————. 2003. *De Crepvscvlis* (1542). *Obras*, vol. II, Henrique Leitão, et al., eds. Lisbon: Academia das Ciências de Lisboa, Fundação Calouste Gulbenkian.

————. 2005. *De erratis orontii: finaei regii mathematicarvm lvtetiae professoris* (1546). *Obras* vol. III, Henrique Leitão, et al., eds. Lisbon: Academia das Ciências de Lisboa, Fundação Calouste Gulbenkian.

ROSSA, Walter. 2002. O Urbanismo Regulado e as Primeiras Cidades Coloniais Portuguesas. Pp. 360-389 in: *A Urbe e o Traço. Uma Década de Estudos sobre o Urbanismo Português.* Coimbra: Almedina.

RYKWERT, Joseph. 1988. *The Idea of a Town.* Cambridge, MA: MIT Press.

TARRIÓ, Ana Maria. 2002. Pedro Nunes e os Humanistas do seu Tempo. Pp. 59-93 in: *Pedro Nunes e Damião de Góis - Dois Rostos do Humanismo Português,* Actas do Colóquio, Aires Augusto do Nascimento, ed. Lisbon: Guimarães Editores, 2002, pp.

TARTAGLIA, Nicolo. 1537. *La nuova scientia (Inventione de Nicolo Tartaglia Brisciano intitolata Scientia nouva divisa in 5 libri, nel Primo di quali si dimostra theoricamente la natura & effeti de corpi egualmente graui: in li dui contraij moti che in esse puon accadere: e de lor cotrarij effeti…).* Venice.

———. 1554. *Quesiti et inventioni diverse de Nicolo Tartaglia. Di novo restampati con una gionta al sesto libro, nella quale si mostra duoi modi di redur una città inespugnabile.* Venice. (Rpt. 1959, Arnaldo Masotti, ed. Brescia: Ateneo di Brescia.)

———. 1556. *La prima [-seconda] parte del General Trattato di Numeri, et Misure di Nicolo Tartaglia, nella quale in diecisette libri si dichiara tutti gli atti operativi, pratiche, et regole necessarie… in tutta l' arte negotiaria, & mercantile, ma anchor in ogni altra arte, sciencia, over disciplina, dove intervenghi il calculo…* Venice.

VITERBO, Francisco Marques de Sousa. 1988. *Dicionário Histórico e Documental dos Arquitectos, Engenheiros e Construtores Portugueses* (1899-1922), 3 vols. Lisbon: Imprensa Nacional - Casa da Moeda.

About the author

Margarida Tavares da Conceição is a historian of architecture and town planning. She earned her Ph.D in Architecture (Theory and History of Architecture) from the University of Coimbra (2008-2009), with a dissertation entitled "On city and fortification in Portuguese texts (1540-1640)". She is a Researcher at the Centre for Social Studies, University of Coimbra. She is also Senior technician at the Portuguese Institute for Housing and Urban Renewal (IHRU), where she has been a member of the SIPA (Architectural Heritage Information System) team since January 2010. Her research interests are: early modern fortified towns; treatises on architecture, town planning and related areas; cataloguing methods for the urban and architectural heritage.

Arturo Lyon*
*Corresponding author

Escuela de Arquitectura
Pontificia Universidad Católica
de Chile
El Comendador 1916
Providencia, Santiago CHILE
alyon@uc.cl

Rodrigo Garcia

Depto. Diseño y Teoria de la
Arquitectura
Universidad Del Bío-Bío
Avda. Collao 1202
Concepción CHILE
rgarcia@ubiobio.cl

Keywords: parametric
constructions, digital
manufacturing, structures, CNC
system

Research

Interlocking, Ribbing and Folding: Explorations in Parametric Constructions

Abstract. This paper presents three projects involving the design and fabrication of architectural structures through the use of different parametric software and digital manufacturing methods. The first project is a flexible partition composed of interlocking elements shaped using a laser-cutter. The second project is a university exhibition unit made with various wooden panels manufactured through a computer numerical controlled (CNC) system. The third project is a system of metal sheets folded by digital machines to create urban circulation spaces. The three works develop a parametric programming of geometry based on certain technical factors, enabling the recognition of patterns of interaction between formal and constructive issues involved in the definition of shapes through parametric controls. Differences in materials and processes are contrasted by similarities of function and conditions involved, creating a system of local, global, productive and environmental parameters that produces a repertoire of self-similar dimensions and variations as well as multiple possibilities of initial setups and final configurations. It suggests a specific field of design exploration focusing on the development of differentiated components and variable architectural configurations, in a kind of open parametric system.

Introduction

The consideration of technical aspects early on in the process of architectural design has traditionally been considered a necessity for developing proper solutions, but also represents the possibility for promoting creative alternatives. Works by Antoni Gaudí, Frei Otto, Eladio Dieste and Santiago Calatrava are frequently cited as key examples where architectural forms have been conceived through innovative structural and constructive considerations. Nowadays the management and performance of building construction has become substantially more detailed, influencing the constraints, products and complexity of buildings. This necessitates a closer relationship between technical conditions and architectural shape, through design strategies that combine them to support creative explorations.

Parametric design software makes it possible to control numerical aspects of the definition of shapes, establishing mathematical relationships between geometry and varying conditions. Several examples of architectural designs have shown the use of parametric software for the integration of different requirements (technical, environmental, functional, etc.) or specific evaluation criteria [Meredith 2008]. The application of this technology during the initial stages of architectural design permits the exploration of multiple possible solutions within determined constrains [Shea et al. 2005]. Nevertheless, no overall design strategy based on these technologies has been fully defined with regard to the relationship between different factors involved or their specific architectural possibilities.

This paper presents three projects carried out by the authors in the design and definition of architectural structures through the use of different parametric design software and digital fabrication methods, reviewing similar conditions in order to identify common design processes and technical aspects involved.

Design parameters can be considered on diverse levels of a project to define particular elements, to control overall form, to control production constraints of elements and design features, and to respond to site conditions that are determinant for several aspects. A conceptual structure of four types of parameters is proposed: production parameters (PP); local parameters (LP); global parameters (GP); and environmental parameters (EP). These categories make it possible to recognize interactions between the factors taken into consideration in the definition of shape in the projects presented.

Project 1: Pixel-Wall

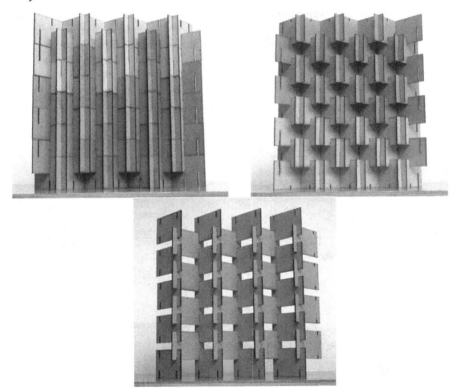

Fig 1. Initial scale-models of Pixel-Wall

Research into a system composed of interlocking plates (called "Pixel-Wall"), was initiated in mid-2009 at the Universidad del Bio-Bio, in Concepcion, Chile by researchers Rodrigo Garcia Alvarado, Underlea Bruscato, Oscar Otarola and Karina Morales.

The system was initially developed on industrial medium-density fibreboard (MDF) boards subdivided into regular elements measuring 61 x 38 cm. (sixteen per panel) to fit within the maximum work area of a medium-size laser cutter. Slots for assembly purposes are cut from these rectangular pieces, to permit interlocking at perpendicular angles. This makes it possible to generate a sequence of interlocked plates which are assembled with other sequences of plates to produce a three-dimensional lattice. Because of the width of the pattern, this configuration is capable of being structurally self-supporting. This arrangement of plates also combines visual enclosure with inner openness, allowing ventilation and indirect illumination, as well as having an expressive appearance and the possibility to hold small objects.

A vertical lattice can be used as partitions or decorative panels. Constructed in more permanent materials, it can be used for outdoor fences, acoustic barriers or ventilated facades. Used horizontally, it can be used for permeable roofs or full envelopes, with varying shapes, transparency, ventilation and noise reduction.

The structural capacity of this configuration is increased by the introduction of curvature in the general configuration, which also allows more formal variety. The inclination of the lateral grooves can create transversal curvature, depending on the angle and length of each slot. In turn, the change in the position of the lateral grooves in the edges can develop a longitudinal curvature (although this stresses the plates). The inclination of the main grooves produces a torque plate that develops curves in both directions. These minor changes in the geometry of each plate result in major modifications in the arrangement. Further, to obtain certain general overall shapes, it is necessary to determine the specific conditions of slots.

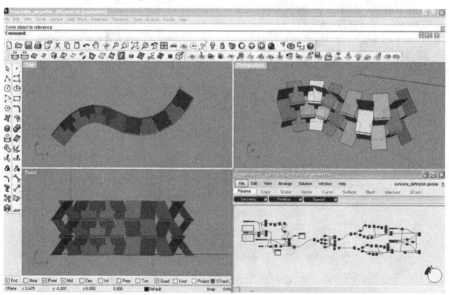

Fig 2. Parametric programming of curved designs of Pixel-Wall

The geometrical relationship between overall configuration and the slots of the panel was implemented through parametric modeling software (Rhinoceros with Grasshopper™). It establish the path of a curve as the axis of the vertical arrangement to develop the plates interlocked according to different amount of pieces in length and height, as well as diverse sizes to regard several materials. The programming makes it possible to generate the complete profiles of the pieces and their respective slots for cutting.

Several full-scale installations of the system have been carried out to test design and execution. For example a decorative partition was installed in the university hall (fig. 3), based on a spline curve 9 meters long and 1.6 meters high, using 160 plates cut from ten MDF boards. This installation was executed in two days and assembled in a couple of hours. The same pieces were used to assembly other configurations, such as curved coverings.

Fig. 3. Installation of Pixel-Wall

Project 2: UMBRALES 2009

The exhibition "Umbrales 2009", held at the Pontificia Universidad Catolica de Chile in Santiago, was a design exercise built at full scale, which was developed in an academic Workshop led by Arturo Lyon and Claudio Labarca. It took as a starting point the definition of specific relationships between the spectators' angles of vision and the variable angles of inclination of the supports for the exhibition of a selection of models and drawings of projects carried out by the students during the previous year.

These relationships were established based on a parametric display module, producing multiple variations in response to changing viewing conditions and to the different kinds of materials exhibited. This system was parametrically modeled using the software Digital Project™, establishing a series of ribs of equal topological conditions, whose angles and dimensions could be adjusted in response to a local geometric mechanism reacting to changes in global parameters. The original module was replicated to produce the entire aggregation, which consisted of 160 different modules and was capable of adapting to changes in the controlling parameters of the overall organization, adjusting their angles and internal dimensions accordingly.

Each rib had to respond to varying angles on the two supported boards, so that it was necessary to include a local scripted reaction to evaluate neighboring supports and determine whether an individual rib was ascending or descending in relation to adjacent ribs, triggering the activation of the corresponding geometrical solution. These changes are translated simultaneously as updates to the manufacturing drawings generated from the three-dimensional model. Each component is composed of seven parts: four segments and three joints. The 1,120 individual parts were cut by a CNC (computer numerical controlled) Router, and then coded and assembled on site.

The exhibition was installed in three different places, with the arrangements changing according the spaces provided, the flow of visitors, and illumination. It used different lengths of series and layouts to create a combination of materials displayed based on the planned exhibit itinerary.

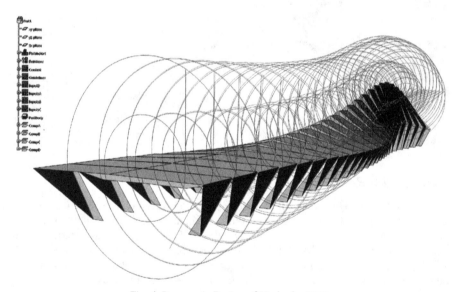

Fig. 4. Parametric Design of Umbrales 2009

Fig. 5. Elaboration of Umbrales 2009

Fig. 6. Installation of Umbrales 2009

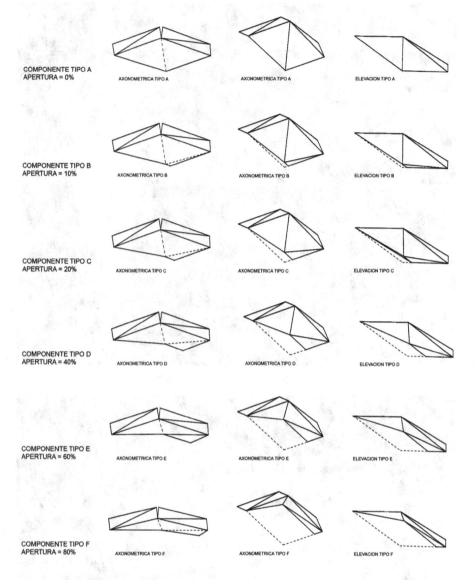

Fig. 7. Parametric design of Plega

Project 3: Plega

The exploration of two architectural projects based on the design of a structural skin system was carried out by Arturo Lyon with collaboration of AKT (Adams Kara Taylor Structural Engineers) and Ivan Valdez in 2008-2009. The skin is composed of aggregations of diamond-shaped components built by folding thin metal sheets (0.7 mm. to 2 mm. thickness), which are cut and folded into rigid three-dimensional panels. The structural quality of the components is provided by resistant edges formed by the folds, reducing the distances of buckling of the sheets while defining a mesh to distribute stresses across a continuous surface.

The Plega system can be adapted to planes or to surfaces of single and double curvature through the use of two-dimensional arrays of parametric components that adjust their internal configuration, while always maintaining the faces coplanar with its four immediate neighbors in order to adapt to non-homogeneous surfaces determined in relation to specific contextual and programmatic conditions. The local variation of the degree of opening of the skin is defined by structural and permeability criteria according to the position of the component in the overall configuration. These local parameters are controlled by data spreadsheets associated to the parametric models developed in the software Digital Project™.

Fig. 8. Application of Plega in a footbridge

Fig. 9. Application of Plega in a pavilion. a, above) exterior view; b, below) interior view

Plega footbridge. In this project the Plega structural skin system was exploited to bridge a span of 40 meters, forming a footbridge that crossed a highway and provided access to a nearby public building (fig. 8). The structural system was designed based on a resistant shell made of a series of fifty inverted arches in the shape of an U. This sequence of components was adapted to the overall defined configuration by integrating specific contextual conditions on the connecting edges of the footbridge, its continuity with preexisting paths and traffic restriction.

Plega pavilion. This temporary pavilion for the Biennale of Architecture in La Paz, Bolivia, was based on a vault made of fourteen interconnected catenary arches with eighteen components each (fig. 9). The proposal at the urban level was to create a shortcut between two busy pedestrian flows, orienting the path between the square and public building that contained the exhibitions of Architecture Biennale. This project was tested through parallel prototypes using CNC machining for cutting and folding of 2mm. thick metal sheets and by printing plans to produce cutting guides for traditional hand tools using 0.7mm. thick metal sheets.

Analysis

Similar conditions ascertained across the three projects described above, developed independently by the authors of this paper in association with different collaborators, make it possible to define common characteristics of parametric design and fabrication processes (Table 1). Although materials (MDF boards, plywood boards and metal sheets), manufacturing methods, processes and dimensions differ, all three projects featured industrialized boards of regular dimensions of about 2 x 1 meters which were manufactured using digital-controlled machines, dividing them into parts smaller than one square meter, with differentiated configurations that allow for the formation of three-dimensional spaces.

The components implemented were variously shaped (rectangular, longitudinal or diamond), and specific details were drawn up to establish up their assemblage (slots, supports and folds respectively). The elements had different configurations, thicknesses and treatments, but in all three cases the individual parts worked jointly to achieve their tectonic qualities as well as their capacity for bearing weight and dividing spaces. The limited structural resistance of these materials, the result of their being mostly intended for decorative use, is augmented by exploring configurations that can result in higher bearing capacity despite their relative thinness (less than 20 mm). They perform by transmitting planar stresses between the elements in different ways (compression, bending and shear) favored by overall forms and connectivity conditions, developing three-dimensional distributions of variable loads, behaving in all three cases as space structures.

The capacity of the components to perform as partitions is mainly a result of opacity and the ability to support lightweight graphic material, which are achieved by the industrial properties of the material defined to fulfill similar purposes. Environmental protection is porous and related to semi-enclosed space, intended as temporary structures that do not need to provide complete protection, since they remain permeable and do not comprise additional elements to achieve acoustic or thermal insulation. Although they performed relatively temporary, non-specific and simple roles, no easy ways to improve these aspects has presented itself, and their nature as an aggregation of individual interlocked elements makes achieving such characteristics complex. On the other hand, they exploit the capacities of materials in their original states, without adding finishes as is usually done, but relying on repetition and continuous variation to define textures through the slight interstices between components. The original treatment is made evident from different viewing angles and light conditions, and through variable volumetric formations, deploying a wide range of spaces with a remarkably efficient use of materials to give a considerable constructive effectiveness.

	PIXEL-WALL	UMBRALES	PLEGA
Element			
Material	*MDF boards*	*Plywood*	*Metal sheets*
Fabrication	*Laser cutter*	*CNC Router*	*Industrial Cutter*
Shape	*Rectangle*	*Lozenge*	*Diamond*
Size	*61 x 38 cm*	*Variable*	*Variable*
Thickness	*6-9 mm*	*12 mm*	*2 - 0.7 mm*
Assemblage	*Slots*	*Supports*	*Folding*
Variations	*Length of slots*	*Inclination*	*Angles & Opening Size*
Finishing	*Reconstituted Wood*	*White sheets*	*Shiny steel*
Composition	*Rows of couples interlocked*	*Ribs of modules with 4 segments and 3 joints*	*Skin of folded pieces*
Design Software	*Rhino-Grasshopper™*	*Digital Project™*	*Digital Project™*
Applications			
Properties	*Partial Opacity*	*Support of Graphic Material*	*Covering or Structural Span*
Functions	*Partitions*	*Exhibition Display*	*Circulation*
Constructions	*Indoor Installations*	*Academic Exhibition*	*Pedestrian Tunnel and Bridges*
Dimensions	*12 x 3 m.*	*2.20 x 80 m.*	*4 x 12 m / 2.5 x 40 m.*
Location	*University Hall*	*Cultural Centre*	*Public Entrance*

Table 1. Comparison of Projects Developed

In all three cases the development process consisted of few operations, although more complex than usual because the differentiated elements required more elaborate design instructions and naming systems, which in general terms allows for a significant reductions of time and management in production. The three projects comprise at least two architectural applications (considering that Umbrales exhibition consists of different exhibition modules, which can also be used in different situations) in which the different versions articulate various spatial configurations while maintaining formal similarities due

---

to their construction system. In this sense the different cases use variable components and different architectural configurations, thus comprising an open system rather than the design and implementation of a particular building.

Elements and designs define formations that escape conventional solutions, especially because of their lack of regular modulation. The projects carried out were temporary and lightweight constructions suitable for experimentation, and did not take cladding, services or internal configurations into consideration. These developments and their results show common conditions that express a methodological procedure and open a new field of architectural possibilities based on digital design and manufacturing technologies.

Parameters

Regarding the categories of parameters suggested earlier, in these cases the local parameters (LP) are mostly the shape and dimensions of the components, which varied to produce constructive connectivity with neighboring parts (relations between components), production possibilities (size of the original elements) and spatial potentials (size of enclosures). To a lesser extent, the materials vary locally, according to properties of resistance and insulation determined industrially, which define quantities and qualities of spaces. Finishes, patterns and size are the main expressive qualities of the configurations.

Global parameters (GP) are essentially the overall dimensions of the enclosure spaces, which determine the amount of components (according local parameters), with a formal subdivision that contributes to general capacities for resistance (curved or diagonal distributions) and takes into account functional capacities (internal sizes). Generating a global spatial formation with semantic, functional and environmental conditions, based on certain outdoor properties as environmental parameters (EP), such as the dimensions of the usable site, orientation, regulation, visual projection, functional relationship, and urban or historical references, to determine the overall characteristics of form.

Production parameters (PP) refer to specific restrictions of dimension involved in the fabrication process, which are determined by the size of industrial products used and the work areas of the different CNC machines used. Along with a repertoire of specific commercial products with different technical capacities and available finishes, this aspect is also affected by the technical skills of those producing the components and carrying out the assembly, and specific circumstances for implementation (transport, setup, etc.).

This combination of factors determined the architectural configurations, some of which were implemented through parametric programming, with a few alternatives tested on physical prototypes, making evident the field of possibilities implicit in the projects and the need to control the formal relations and techniques involved.

A conceptual scheme of parameters at the different levels is given in Table 2. This diagram is neither exhaustive nor precise, but helps illustrate how interaction between several characteristics of work can be managed in parametric design. In production parameters (PP) there are many physical properties of material or manufacturing conditions. Local parameters (LP) regards the piece manufactured, and usually involves defining several features based on the physical properties of the materials combined with the geometrical dimension or shape. Global parameters (GP) regard the main characteristics of the design, usually based on sections or group of components, and several interior properties based on features of the elements related to the space configured. Environmental parameters (EP) regard some conditions of site which constrain total volume, situation and features. Many of these features are not easily converted to numeric factors or geometrical relationships, due to the use of different units or the presence of complex conditions. However, it allows us to see how some specifications can be managed to establish an open design system.

PP- Production Parameters	LP- Local Parameters	GP - Global Parameters	EP - Environmental Parameters
Dimensions of industrialized material			
Dimension of working area in manufacturing machine	Dimensions of component	Section size + sequence	
	Shape	Dimensions of complete design	Dimensions of site
	Connection System	Shape	
Colour		Function	Activities in surrounding
Roughness	Appearance	Texture	Expression
	Opacity	Visual relationships	Views
Strengh	Resistance	Structural stability	Geological conditions
Acoustic transmission	Soundproofing	Sound level	Sound sources
	Openness	Ventilation	Wind direction
		Illumination (or Luminance)	Orientation
Conductivity	Insulation	Indoor temperature	Climate

Table 2. Parameters

Conclusions

This paper has described three projects of architectural arrangements with different components developed through the use of parametric design and digital manufacturing. The analysis of their properties reveals material and formal differences, but significant similarities in the organization of result, processes and products involved, comprising a structure of local, global, productive and environmental parameters where different aspects involved can be recognized. We also acknowledge that implemented parameters correspond to conditions related to manufacturing, connectivity, support, functionality and expressivity of the elements and the proposed form, presenting a repertoire of similar dimensions and variations as well as sets of initial products and final configurations, suggesting a specific field of exploration focusing on the kind of lightweight installations outlined. In this sense it would be relevant for further research to focus on aspects related to more complete conditions, such as structure, space and insulation.

The projects suggest the development of design systems based on some of these implemented parameters in various computing platforms defining open and variables processes, where the configurations developed appear as circumstantial possibilities. This suggests a particular kind of architectural research, aimed at defining general procedures, with control of specific elements and conditions, extracting specific solutions for specific situations, unlike the conventional process, which starts with a wide range of initial elements and combinations to execute specific works. Following this approach, buildings are set as combinations determined by the intensive application of technical aspects.

Acknowledgments

This work is part of research project FONDECYT 1100374.

References

AMBROSE, M., CALLAM B., KUNKEL J. and WILSON L. 2009. How To Make A Digi-Brick. Pp. 5-12 in *Proceedings of the 14th International Conference on Computer Aided Architectural Design Research in Asia*, 22-25 April 2009. Yunlin (Taiwan): National Yunlin University of Science and Technology.

ARPAK, A., L. SASS, and T. KNIGHT. 2009. A Meta-Cognitive Inquiry into Digital Fabrication: Exploring the Activity of Designing and Making of a Wall Screen. Pp. 475-482 in *Computation: 27th eCAADe Conference Proceedings*, 16-19 September 2009. Istanbul (Turkey): Yildiz Technical University.

BONSWETCH, T., D. KOBEL, F. GRAMAZIO, and M. KOHLER. 2006. The Informed Wall: Applying additive digital fabrication techniques on architecture. Pp. 489-495 in *Proceedings of the 25th Annual Conference of the Association for Computer-Aided Design in Architecture*, Louisville (Kentucky) University of Kentucky, Lexington.

BROOKES, A. J., and D. POOLE. 2004. *Innovation in architecture.* London: Spon Press – Taylor & Francis Group.

CALDAS, L. and J. DUARTE. 2005. Fabricating Ceramic Covers. Pp. 269-276 in *23nd eCAADe Conference Proceedings*, 21-24 September 2005. Lisbon, Technical University of Lisbon.

KIERAN, S. and J. TIMBERLAKE. 2004. *Refabricating Architecture, How Manufacturing Methodologies Are Poised to Transform Building Construction.* New York: McGraw-Hill.

MENGES, A. 2009. Performative Wood: Integral Computational Design for Timber Constructions. Pp. 66-74 in *Proceedings of the 29th Annual Conference of the Association for Computer Aided Design in Architecture (ACADIA)*, , 22-25 October, 2009. Chicago: The School of the Art Institute of Chicago.

MEREDITH, M., ed. 2008. *From Control to Design: Parametric/Algorithmic Architecture.* Barcelona: Actar.

OESTERLE, S. 2009. Cultural Performance in Robotic Timber Construction. Pp. 194-200 in *Proceedings of the 29th Annual Conference of the Association for Computer Aided Design in Architecture (ACADIA)*, , 22-25 October, 2009. Chicago: The School of the Art Institute of Chicago.

SASS, L. and M. BOTHA. 2006. The Instant House:A Model of Design Production with Digital Fabrication. *International Journal of Architectural Computing* 4, 4: 109-123.

SHEA, K., AISH, R., and GOURTOVAIA, M. (2005). Towards Integrated Performance-driven Generative Design Tools. In *Automation in Construction* 14, 2: 253-264.

STACEY, M., ed. 2004. *Digital Fabricators.* Waterloo (Canada): University of Waterloo School of Architecture Press.

ZISIMOPOULOU, K. and A. FRAGKIADAKIS. 2006. Constructing the String Wall - Mapping the Material Process, Communicating Space(s). Pp. 326-335 in *24th eCAADe Conference Proceedings*, 6-9 September 2006. Volos (Greece): University of Thessaly.

About the authors

Arturo Lyon is an architect who graduated from the Catholic University in Chile and Masterin Architecture from AA's Design Research Laboratory. After finishing his studies he worked at Zaha Hadid Architechts in London for two years focusing on the use of parametric design and algorithms for the resolution of doubled curved paneling systems as well as conceptual designs for different projects in China and Middle East. He has taught at the AA Landscape Urbanism Programme and workshops on Generative Algorithms and Parametric Design at the AA and at the Catholic University in Chile. He is currently Assistant Professor at the Catholic University in Chile and partner at LyonBosch architects.

Rodrigo Garcia has a Ph.D. in architecture from Universidad Politecnica de Catalunya, Spain (2005), Master in Information Technologies for Architecture from Universidad Politecnica de Madrid, Spain (1994) and Bch. in Architecture from Pontificia Universidad Catolica de Chile (1989). He has been visiting researcher in University of Kaiserslautern and Bauhaus-Univeristy of Weimar, Germany, University of Houston, USA and Stratchclyde University, UK. He works on digital media in architecture, in particular 3D modeling, animation and digital fabrication. He also carries out research on design teaching, contemporary architecture and virtual heritage. Currently he is head of the Ph.D. programme in Architecture and Urbanism at the Universidad del Bio-Bio, Concepcion, Chile.

Mustapha
Ben-Hamouche

College of Engineering
Department of Architecture and
Civil Engineering
University of Bahrain
P.O. Box 32038
Isa Town, Kingdom of Bahrain
mbenhamouche@eng.uob.bh

Keywords: Islamic law,
succession, urban design,
subdivision, iteration, fraction,
fractals, chaos, morphology,
urban geometry

Research

Fractal Geometry in Muslim Cities: How Succession Law Shaped Morphology

Abstract. Islamic succession law has deeply affected the urban fabric of Muslim cities. Properties were subdivided according to a refined and elaborate system of shares that were prescribed by jurists. Successive iterations of subdivision over the course of decades or even centuries gave a fractal character to the cities and thus becomes the main source of their complexity. Most property was subject to successive subdivisions until minimum but functional parts were arrived at. However, this fragmentation should not be understood in isolation from other mechanisms which sometimes resulted in a reunification of the fragments, and thus established a dynamic equilibrium in the urban fabric. The paper will present the mechanisms of subdivision according to laws of succession, and illustrate by hypothetical example their direct impact on the urban morphology, as a means of understanding the complexity of old Muslim cities.

Introduction

The urban fabric of Muslim cities has been an enigmatic issue for scholars. Early studies of orientalists concluded that Muslim cities were lacking order due to the absence of institutions and the excess of freedom of action that characterized their communities. The absence of straight lines, the interlocking of buildings and encroachments on public realm were seen as evidence of this disorder. According to this view, this was mostly due to the laxness of the authorities and officials, who were mainly interested in harems and holy wars [Sauvaget 1934; Letourneau 1985; Raymond 1994].

It was only in the middle of the twentieth century that a new generation of scholars changed the paradigm of research, finding in Islamic law and institutions determinant factors for this complexity [Barbier 1900; Brunschwig 1947; Hakim 1988]. Recent studies showed that the irregularity of geometry and urban fabric in these cities is not a sign of disorder and chaos in the pejorative sense, but rather a sign of high order and complexity [Akbar 1988; Ben-Hamouche 2004, 2009a, 2009b].

Chaos and fractal theory provides an efficient tool for analyzing the urban fabric of old Muslim cities. It presents a new instrument for understanding the complexity of these cities, and thus displaces the Cartesian approach and Euclidean geometry that have long dominated the studies of urban fabric and morphology.

Succession law intervenes at turning points in life, that is, death and birth. Social interactions, regeneration and shifts of property provide an opportunity for the application of the theory of chaos and fractals to the domain of social science. However, this present study will be limited to the causal relationships between succession as a field of Islamic law and the morphology of urban fabric in Muslim cities. Such relationships are believed to be a major factor behind complexity in Muslim cities. A symbiosis of arithmetic for the calculation of heirs shares and geometry for the subdivision and partition of assets has generated the urban morphology of these cities; without taking this into consideration, any analysis of their urban fabric would be a mere *tatonnement*.

The aims of this study are to contribute to the advancement of the theory of chaos and fractals through an enlargement of its scope, and to provide a new instrument for academicians and professionals for understanding and analyzing the urban fabric of old Muslim cities. Such a contribution would lead to a shift in academic fields of research, as well as education regarding the study of these cities. On a practical level, reconsidering succession law in urban regulation and municipal systems would have direct implications for the urban management and preservation of these cities.

Fractals: back to the roots

The terms "fractals" and "chaos" are organically interwoven, and together form a single, integrated theory. They are usually used concomitantly [Devaney 1990; Briggs 1992; Szemplinska-Stupnicka 2003]. They are concerned with irregular forms and complex objects that show self-similarity at each scale of magnification when they undergo a certain process of iteration. However, they sometimes have distinct connotations and applications in different fields of knowledge. Consequently, the present study is limited to the geometric aspect of the theory.

Literally, the term fractal, coined by Benoit Mandelbrot in 1975, was derived from the Latin *fractus*, meaning "broken" or "fractured". Mathematicians developed fractal dimension D to measure self-similarity through scale invariance. It is defined as the power relation between the number of pieces and the reduction factor: $A= 1/(s)D$. Fractal dimension D could thus be defined as: $D = \log(a)/\log(1/s)$ [Bechhoefer and Bovill 1994: 27].

In this study fractal geometry does not entirely adhere to this conventional meaning, nor to the instruments and measurements rules as used in many previous studies [Bovill 1996; Bechhoefer and Appleby 1997; Eglash 1999; Ostwald 2001]. However, it shares the same philosophy and roots of the term. In technical terms, it is related to the arithmetic meaning of "fraction", that is, a quotient number that represents part of a whole. Applied to succession law, it reflects the endless process of fragmentation of inherited properties into smaller shares and the resulting effect on the urban fabric. This (re)definition is not necessarily opposed to the essence of the theory, but broadens its scope. Bypassing its present technical aspect, which is imprisoning it in mathematical formulae, would nurture its philosophy and help its diffusion.

The scope, source and nature of Islamic succession law

Islamic succession law is believed to lie at the basis of the birth and development of both algebra and geometry in Islamic civilization. Al-Khawarizmi (780-850 A.G.), from whose name the term "algorithm" was derived, introduced algebra to modern mathematics through examples from Islamic inheritance laws [Huma 1997]. In geometry, techniques and tools for the calculation of areas on the ground and shares of heirs were also made available to jurists and dividers [Cami-Efendi 2000; Ca'fer Efendi 1987].

Measuring systems, calculation methods and geometric techniques were developed in Islamic civilization as a result of the pressing need to partition and calculate irregular shapes. In other words, developments in mathematics, and in particular in algebra, were a direct consequence of Islamic laws of inheritance [Cami Efendi 2000: 117-118, 123-

130]. A body of hypothetical science involving tricks and cunning was also developed as an entertaining branch of mathematics [Cami Efendi 2000: 131-134].

Socially, Islamic succession law is deeply rooted in Muslim communities. To colonial authorities during the nineteenth century, it was one of the key themes for understanding social structure and managing administrative affairs of the local societies. Many manuscripts were translated into English and French, the languages of the two major ex-colonial powers, in order to ease the mission of their administration. *Hedaya* is an example of a translated manual that was concerned with generalities in Islamic law [Hamilton 1989], while the *Al-Sarajiyya* manual [Rumsey and Jones 1869], on which most jurists of the Hanafi school of thought in the Indian continent relied for succession law, was one of the famous documents that was edited by the British lawyer Almaric Rumsey (1825-1899).[1]

Islamic inheritance law is considered to be a branch of jurisdiction. It is a sole prerogative of a judge, *Qadi*. An unpublished manuscript that goes back to the eighteen century, recently annotated and edited by the present author, on which part of this study is based, is *Riyadh al qassimeen*, the "Garden of the Dividers" [Ben-Hamouche 2000]. The author, Cami Efendi, who lived between 1649 and 1723, was an Ottoman judge who used it as a manual for the daily affairs of partition and urban disputes in courts of Cairo and Bursa, Turkey.

Methods of partitions and rules applied to movable objects as well as real property. According to jurists, any property that belonged to a deceased person is considered to be an inheritable asset to be subdivided, no matter how small or large. However, many objects such as wells, mills, wind-towers, ovens, and small pools were considered to be undividable. This was also the case of objects that would become useless or not functional after partition [Cami Efendi 2000: 51-60]. Passages, and roads, courtyards and staircases were subject to this same principle.

A partitioning of usufruct, called *Muhayat*, was applied in the case of properties that could not be subdivided, due either to the large number of heirs, or the small size, or the nature of the object, as will be seen in sections below. Another alternative to physical partitioning of properties would simply be to sell the inherited property and subdivide the income according to the same rules of subdivision.

As a principle, it is recommended that after subdivision each partner enjoys his share separately, with his own road and drainage. Otherwise, the partners must agree to enjoy their respective shares with all rights and immunities in common. If such conditions are not met, the partition must be annulled and made anew [Hamilton 1989: 572].

After subdivision, and according to the number of the heirs, a rule of thumb is applied to determine the share of each heir for the sake of impartiality and pleasure. However, the partition is acceptable if the heirs reach a consent without the intervention of authorities. Once the partition is determined and each member receives his share, it cannot be cancelled.

Islamic succession law is often criticised because of its position regarding women, who are seen as disfavored. A sister, for instance, is given half the share of her brother from the estate of her deceased father. This is, however, only one social case among others. In some circumstances, she receives a share equal to that of a man, as in the case of the brothers and sisters of a deceased man. In some others, she has more than the males' shares [Rumsey and Jones 1869: 6]. In other cases she gets a share of a subscribed heir while her male counterpart is entirely excluded from the partition. On a large scale, the traditional social system was organized so that males had financial duties towards children, wives and parents. Females were entirely exempted [Yekini 2008: 56-60; Abdul-Jamal 2005].

Another criticism made of the system is its impact on the agricultural land, claiming that Islamic succession law leads to its being fragmented and thus inefficient in terms of productivity and economy of scale [UN-Habitat 2005: 9]. This vision, mostly based on the pro-capitalist standpoint, seems to attribute the weak economy of less developed countries to land tenure through an unfair comparison between rich and poor countries, and disregards other less visible problems underlying economic weakness. It also contests the distributive philosophy behind the succession law, which aims at socializing means and capitals and preventing monopoly and the accumulation of wealth in the hands of a few rich people. Redistribution of material resources, including real estate, is the core of the modern economy and a key concept for social justice, offering equal opportunities to access wealth and means to wealth [De Soto 2000].

The mathematics of inheritance

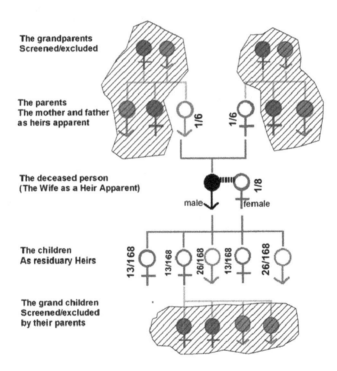

Fig. 1. An example of a family structure of a deceased man and the share of the heirs according to their kinship

Islamic law of succession is fundamentally a combination of arithmetic, sociology and geometry. Shares are well defined according to kinship and family relationships. On the basis of a few fundamental texts, jurists of different schools of Islamic law developed a complex system of inheritance that is largely applied in most Muslim countries today.

First the heirs are determined, then complex methods of calculations are applied in theory. At a third stage, assets such as houses, stores and lands are subject to a process of subdivision. The sections below will outline the system, enumerate the heirs, and explain the way the assets of a deceased person are subdivided among them (fig. 1).

The heirs apparent

The first class of heirs apparent or sharers, called *As'hab al furudh*, comprises the circle of relatives closest to the deceased defined explicitly in the fundamental texts [Rumsey and Jones 1869: 4-7]. The shares, depending on their family status, are defined in terms of fractions to which they are entitled. Fractions are multiples of either the 1/2 or the 1/3. The ones stated in the *Quran* are the 1/2 , the 1/4 , the 1/8, the 1/3, the 2/3 and the 1/6 (Table 1). Heirs entitled to these shares are twelve persons; four males and eight females. In the first subclass we find the father and the true grandfather or other male ancestor, however far up in the paternal line, and the brother by the same mother and husband. In the second subclass we find the wife, the daughter and the son's daughter or other female descendant however far down, the sister by one father and mother, the sister by the father's side, and the sister by the mother's side, the mother and the true grandmother, that is, she who is related to the deceased without the intervention for a false grandfather[2] [Rumsey and Jones 1869: 4; Yekini 2008: 43].

The exclusive presence of this class makes subdivision a matter of a simple arithmetic calculation that consists in defining the shares as percentages of the sum total of assets. Often a lowest common denominator is found for this purpose. The daughter who is the only heir receives 1/2 of her father's estate. A wife is entitled to 1/4 of the property of her deceased husband if he does not leave children, and only 1/8 if he does. The husband receives 1/2 the property of his deceased wife if she doesn't leave any children, and 1/4 if she does.

However, if within the class of similar heirs, such as many daughters and/or sisters, the share is further subdivided equally among them. For instance, wives of a deceased person in the presence of children from him would share the 1/8 among them. If only girls are left and number two or more, they are entitled to 2/3 of their father's property. This share is further subdivided equally among them equally. For instance, if there are five daughters, each is entitled to 2/15 of the initial share, which was 2/3 of the estate. The remnant, i.e., 1/3, is divided among the other relatives. Table 1 gives the list of shareholders classified according to fraction of share.

In some cases, a partition among the primary heirs may lead to a residual part. For instance, the estate of a deceased person who has two sisters (2/3) and a mother (1/6) will lead to a surplus part of 1/6. Another process of partition, called return or *radd*, is then applied to the subdivide the remaining 1/6, where each primary heir is entitled to a new portion of the residue in proportional to his or her prescribed fraction. The initial share is then altered after adding a new portion of the remaining part. Other jurists stipulate that the remaining part is to be taken by the public treasury [Rumsey and Jones 1869: 27].

Conversely, sometimes the total sum of the assigned shares of the heirs becomes greater than unity. A subtraction process, called *al-awl*, is then applied. A deduction is made from each share in proportional to its relation to the whole so that the deficit is divided fairly. For instance, a deceased woman may leave competing primary heirs, such as a husband who is entitled to 1/2, two full sisters who are entitled to 2/3, and a mother who has a right to 1/3 of the property. In this case, since the sum of the shares is greater than the whole estate, all the shares are reduced proportionately. The fractional shares are consequently changed to a new common denominator, that is equal to the sum of the numerators (Table 2).

#	The basic fractions	The sharers*	Conditions
A	1/2	1-The husband, 2-the only daughter, 3-the son's daughter, 4-the sister, 5-the father's sister.	1-if the wife doesn't have offspring 2-if she has no brother or sister 3-if there is no other son or son's son 4-if there is no brother, father, son or son's son. 5-if there is no father's brother, and as in (4)
B	1/4	1-The husband, 2-the wife (wives)	1-if the wife has no offspring 2-if the husband has no offspring
C	1/8	1-The wife/widow (wives)	1-if the husband has offspring
D	2/3	1-Two daughters or more, 2-the two sisters or more, 3-the two father's sisters or more, 4-the two son's daughters.	1-if there is no son 2-if there is no brother, son(s),son's son or father 3-if there is no father's brothers plus the conditions above (2) 4-if there is no son or son's son.
E	1/3	1-The mother, 2-the brothers' mother, 3-the grand-father.	1-if there is no son, son's son and brothers 2-if they are numerous and there is no father, no grand father or son and no son's son 3- If there are brothers and the 1/3 is the highest than in any other alternative subdivision
F	1/6	1-The mother, 2-the grand mother, 3-the father, 4-the grand-father, 5-the mother's brother, 6-the son's daughter, 7-the father's sister.	1-if there is a son, a son's son, or two or more brothers 2-if she is unique (otherwise grand-mothers share the 1/6) 3-If there is a son or son's son 4-If there is a son, a son's son and no father 5-if he is unique, and no son, son's son, father or grandfather 6-if there is only one daughter without her brother, nor her uncle's son 7-if she has a sister and has no brother or father and there is no son's son.

Table 1. The prescribed shares of heirs apparent according to their status and under some conditions. Source: adapted and translated from [Amrani 2000: 62-63]

	Heirs	Old share	New Share
1	The husband	1/2.= 3/6	3/9
2	The Two sisters	2/3.=4/6	4/9
3	The mother	1/3.=2/6	2/9
4	Total	(3+4+2)/6=9/6	9/9

Table 2. Applying the method of deduction on prescribed shares

The residuary heirs

The second class mainly comprises the male agnate relatives to the deceased, known as *Ossba*. They are heirs whose right to inherit could be lost through the birth of a nearer relative. They are of three kinds: the upwards relatives (i.e., the ancestors however far up the line), the downwards relatives (i.e., the sons and the son's sons however far down the line), and the lateral relatives of the deceased and his parents (i.e., the brothers and uncles). In general these heirs have no subscribed shares as do those in the first category; their shares are known only after the first class heirs have taken their prescribed shares. If these latter are deceased, the heirs presumptive have the entire estate.

An order of priority, based on a principle of exclusion called *Hujb*, which literally means screening and/or excluding, is applied to this class and the class below. The aim is to give the closest relative priority over the other heirs, who in this case are partially or totally excluded. A simple form of screening consists of an heir who is excluded due to the presence of other, higher heirs. For instance, a father fully screens the grandfather, and the son fully screens the grandson. However, the screening could be partial, as in the case where heirs are shifted from a position to a lower one. If a deceased man leaves brothers and a mother, the mother will pass from the share of 1/3 to 1/6 due to their presence.

There are various ranks of *'asib*, a person who has no prescriptive share. The first category is the sonship (i.e., son, grandson, etc.), next comes the fatherhood (father, grandfather, etc.), then brotherhood (full and paternal brothers, and their sons), and finally, unclehood (germane uncle, consanguine uncle, and their sons) [Yekini 2008: 50]. According to rules of priority for inheritance, Islamic law also installs a system of partial and full exclusion. For instance, the existence of a higher rank *'asib* cuts off the heirship of those of lower ranks. A son cuts off brothers, while a father cuts off uncles, whereas a son does not cut off a father [Yekini 2008: 50].

In the presence of male and female relatives such as sons and daughters, the rule that applies to the subdivision of inheritance among them is that the males are entitled to double the share of the females. For instance, a son inherits a share equivalent to that of two daughters, a full (germane) brother inherits twice as much as a full sister, a son's son inherits twice as much as a son's daughter, and so on.

In some cases, an overlap occurs between the first and second classes. An heir who is in a given category in the presence of some other heirs, could be shifted to another category in their absence. For instance, if the deceased left only one daughter or agnatic granddaughter (the daughter of the son), her share is fixed as 1/2. However, the presence of the deceased's sons (i.e., her brothers) would shift her to the second category, where she is entitled to a share half that of her brother. If there are two or more daughters or agnatic granddaughters without brothers then their share is two-thirds. Two or more daughters will totally exclude any granddaughters. If there is one daughter and an agnatic granddaughter, the daughter inherits 1/2 and the agnatic granddaughters inherit the remaining 1/6, making a total of 2/3. If there are agnatic grandsons amongst the heirs, then the principle that the male inherits a portion equivalent to that of two females applies.

The distant kindred

This third class consists of larger circle of relatives. It comprises the descendants of the deceased, like the daughters' children, the upper grandmothers and grandfathers, the sisters' and brothers' descendants, and the parental aunts and uncles by the same mother only, as well as maternal uncles and aunts [Rumsey and Jones: 34-35]. Schematically with regards to the deceased, they could be organized into the roots, if directing upwards, the branches if directing downwards, and the far lateral relatives on both sides.

Not all jurists agree about this class in the subdivision. Some maintain that the distant kindred are not entitled to any share at all, as they are not explicitly mentioned in the fundamental texts. According to this opinion, in the absence of the previously stated heirs the property of the deceased is transferred directly to the public treasury. In some cases, the public treasury becomes a member of the heir classes of the deceased. This opinion has also an effect on the process of subdivision of the property.

The importance of such a class depends on the two previous ones. In the presence of the two previous ones, nothing is left to the heirs of this class. However, cases where the two classes disappear are not exceptional. History shows that masses of people died due to epidemics and earthquakes, leading to the extinction of entire families who resided in the same city or a region, and survival of their far relatives in other areas [Ben-Hamouche 2009b: 94-96].

In the majority of cases, most subdivision processes are limited to two or three generations covering a time span of, say, from 60 to 120 years. However, jurists elaborated methods and legal opinions that cover up to four generations. There are exceptional cases, such as a deceased who leaves two cognate far grandfathers from both his mother and father who are the only remnant heirs. The estate in this case is subdivided equally between them (fig. 2).

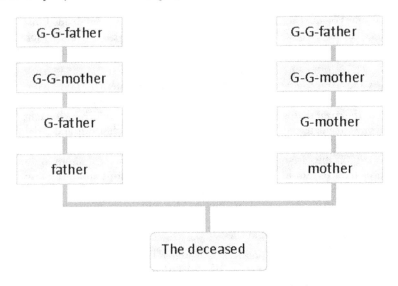

Fig. 2. Example of upwards distant kindred

The presence of many overlapping generations of distant kindred in the absence of the two previous classes leads to a very complex combination of rules that define each heir's share, as highlighted by Rumsey and Jones [1869: 36-48]. The results of the subdivision may turn into a series of many fractions with large denominators, ranging from tens to thousands, depending on the number of heirs.

The logic of subdivision could be mathematically modeled and turned into a simple software. The three classes of heirs as defined above would be the major stages of decision-making in an algorithm that helps lawyers as well as the layman define the shares of each heir depending on the data entered.

The arithmetic that comes from the algorithm could then be displayed graphically through the simultaneous representation of all shares in shapes within a square of 100×100 pixels using the cellular automata system that reflects the stages of iteration.[3]

The geometry of subdivision: uncovering the mystery of complexity

While mechanisms of subdivision acted continuously on the morphology of the city in congruence with the life cycle of social entities (i.e., family units and individuals), they did so together with other mechanisms in an integrated system. When segregated, each one appears to have a negative effect on the morphology. For instance, the pre-emption principle, *Shufa'*, gives the partner in a property or the neighbor in an adjacent property the right to get back the shares of other partners when these latter put them up for sale on the market. Consequently, this mechanism partially acted to counter succession, as shares were unified in one new property. Other mechanisms, such as *waqf, Irtifaq, Ihiya al-mawat,* have similar effects of countering the fragmentation of properties [Ben-Hamouche 2009a, 2009b].

Also, a subdivision is not always the sole or compulsory recourse for partners. A partition of usufruct, *muhayat,* as seen earlier, is an option open to the heirs as long as they reach an agreement. The asset in this case is kept physically intact, and each heir enjoys the use of the thing held in partnership either by turns or simultaneously, where this is possible. For instance, a courtyard could be shared as a subdivided asset, where each partner has his own access and rooms. This is also the case of small shops, public baths, houses and cultivable land [Cami Efendi 2000: 94-101; Hamilton 1989: 576-578].

Passages and roads that are collectively owned may be subjected to subdivision in case of a dispute among the partners, and if this is practicable. If partners disagree regarding the extent of the road, that is regarding the height and breadth which each ought to have, the judge must regulate their proportions according to the breadth and height of the doors of their respective houses, as that is sufficient. According to [Hamilton 1989: 572] the advantage of this arrangement is that if any of them is desirous of making a projection or terrace from his house over the road he may do it above the height of his door but not below it, and the road will still remain in common. Quite often territories of houses along a road are marked by physical elements, such as arches and signs, that have unique architectural significance, but appear as mere aesthetical elements to unknowing visitors, tourists and outsiders.[4]

Basically, an asset, depending on its size as well as the number of the heirs, should be judiciously subdivided in response to the legal shares in terms of areas, as well as autonomy and conditions of use. "Conditions of use" mainly refers to access, drainage of rain water, and provision of light.

In contrast to what appears to be the basic unit in the city, that is, the house, for succession law the spatial unit in a subdivision is the room, or *bayt,* defined as the single roofed place surrounded with walls, with a door or entry. A tenement that is composed of different rooms, a roofed court, and a kitchen, so that a man may reside there with his family, is called *manzal.* This is also the case of the tenement with an open court, termed *dar* [Hamilton 1989: 572]. A direct impact on the morphology of such a derivative concept of domestic space is the setting of property lines after each new subdivision. Following the construction of walls and the lines of interior partitions, only rooms and other undividable elements appear to maintain their integrity in the urban fabric. Two adjacent rooms that once both opened onto the same courtyard may, after partition, have opposite entrances, each of which opens onto a different courtyard. Understanding this may make it possible to decipher the dynamics and complexity of urban geometry at the domestic level in old Muslim cities. By way of recurrence, many houses, after removing the lines of subdivision, could be reconstructed into their initial state, in which, for instance, they had larger courtyards and simpler shapes as blocks, and were surrounded by larger roads.

Winding streets, cul-de-sacs and dead-end streets are often enigmatic urban elements. A portion of land that is located in the heart of the initial property would necessitate a dead-end street and/or a narrow winding alley to provide access, a feature that forms an important portion of the road network in old Muslim cities [Raymond A. 1994]. Consequently, it is evident that a dead-end street and a winding alley that were mainly described by Western scholars as signs of an *a priori* geometry and voluntary irregularity in Islamic cities, are in fact *posteriori* events that resulted from a realistic response to an existing situation.

The partitioning of many houses and tenements located in different places is a source of many controversial legal opinions. The divergence stems from differences in the market value of the property due to the difference in location in the city (or cities), and to different functions, such as shops, stores, gardens, or houses. For instance, if a deceased man left several houses held in partnership or coparcenaries in one city, the grand jurist Abu Haneefa maintained that, for the sake of fairness, each house must be separately divided among the heirs [Hamilton 1989: 570]. For him, the value of the houses may differ greatly depending on the cities, lanes or neighborhood in which they are situated. Their proximity to or distance from water or a mosque makes it impossible to achieve equality in the partitioning without dividing each house separately.

However, other jurists, with a more realistic view, hold that if it is convenient, the whole of the houses would better be united in one general partition and not divided separately. The share of each partner must be assembled as far as possible in one of the houses, so that it may be solely his property. Implicitly, a system of compensation would be devised in order to cover the differences of market value between the different assets.

Complicated partitioning can also occur in the case of buildings with more than one floor. In a case of a house with two floors, the same school of law has three diverging opinions regarding the value of each floor. Abu Haneefa maintains that the ground floor has a value double that of the upper floor, a fact that affects the subdivision. The subdivision into two equal parts of a property with two floors, considering the value of a square meter of the lower floor to be twice that of the upper one, is shown in fig. 3.

Fig. 3. The subdivision of a two-storey house, where the value of the lower floor is twice that of the upper one

The impact on urban geometry

The sections below show the method of calculation of shares, the subdivision process and the impact on the urban geometry. The first one will present the method applied to a simulation case, whereas the second one will present evidence on the ground through a semi-simulation case of an urban fabric in old Algiers.

An application to a hypothetical case[5]

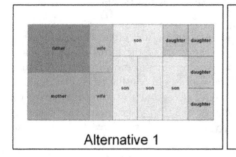

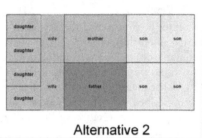

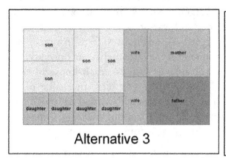

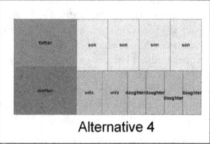

Shares of Heirs and Alternatives of subdivision

Table of Shares

The Inheritors	The shares in fractions	The fractions with l.c.d.	The area in m² X=(60x120)
The mother	1/6	48/288 X	1200
The father	1/6	48/288 X	1200
Wife 1	1/16	18/288 X	450
Wife 2	1/16	18/288 X	450
Each Boy	26/288	26/288 X	650
4 Boys	4*26/288	4*26/288 X	2600
Each Girl	13/288	13/28 X8	325
4 Girls	4*13/288	4*13/288 X	1300
Total		288/288 X	7200

Alternatives designed by Salmane Ben-Hamouche

Fig. 4. A hypothetical subdivision of a land parcel of 60 x 120 m². Alternatives and shares displayed graphically: blue for father, purple for mother, green for wives, yellow for sons, and brown for daughters

A hypothetical property that devolved on a number of heirs best explains the methodology and the process of subdivision and the way fractal geometry is generated. A deceased man leaves a piece of land measuring 60 x 120m² that is accessible from four sides. The asset had devolved on his heirs, who are his two parents, his two wives, and his children, who are four boys and four girls. The subdivision of the property considers the two classes of heirs. The first consists of the sharers who have predetermined parts, and the second is that of the children, who will subdivide the remaining share according to the rule that males have twice the share of females (fig. 4).

The system of subdivision may be further complicated if the frontage elevation is taken into consideration during the operation. A portion of land that gives onto a street would have a higher commercial potential than the one located inside the block or on a dead-end street. The frontage could then be quantified in terms of metrics and subdivided in proportional to the shares, a fact that further increases the complexity of calculations and geometry.

Fractals, iteration and bifurcation

Subdivision is initially a collective decision-making process that consists of agreeing, eliminating alternatives, and maintaining the most convenient one. In addition to his legal status, in this case the divider person would also be a witness and arbitrator. After having defined the shares through calculations, many possibilities figure on the ground to the divider as well as the heirs. Depending on the size of the property as well as the number of shares, a decision is first made about the best alternative of subdivision, one that considers fairness as well as conditions of autonomous use of each part. It is only after this step of agreement is reached that shares are distributed among heirs. Specific allocation of shares to each heir is then made either through another agreement or by drawing lots.

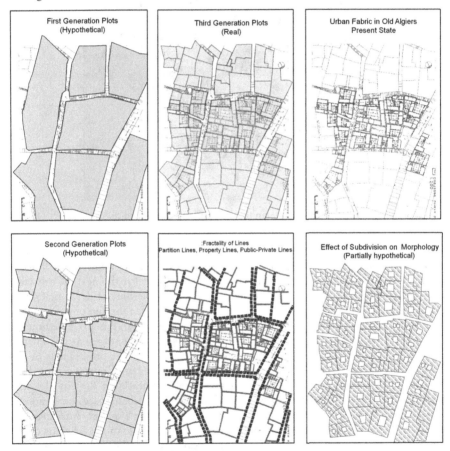

Fig. 5. The impact of succession law on the urban geometry in old Muslim cities. Fractality in lines of urban fabric and the process of land subdivision

Fig. 6. The impact of succession law on the distribution of water and the morphology of desert cities, Tamentit, South of Algeria

According to the Islamic succession law, subdivision is also an iteration process *par excellence*. Iteration in mathematics refers to the process of applying a function repeatedly, using the output from one iteration as the input for the next. When taken over a long time span, say decades and centuries – that is, the time of development and growth of cities – it means an act of repeating an operation with the aim of defining new shares with a combination of prescribed and deduced fractions. An initial simple function of iteration consequently produces complex morphology at later stages, as is known in chaos theory [Batty 1991, Ben-Hamouche 2009a]. The scale of the process may be best realized when we consider the urban dimension and the number of operations made on the daily basis, which is, ironically, related to the rate of deaths in the city. In other words, private assets that form the largest number of the urban space are almost all subject to subdivision. One exception is made for the few public buildings and endowments assets, *waqf*, that are protected from alterations due to the religious clauses of donators. In terms of fractal geometry, a subdivision process is continuously a point of bifurcation that gives rise to many tree-like possibilities of fragmentation, that all heirs establish. However, while in theory such an iteration of subdivision is endless, in reality it stops when it reaches a certain threshold of minimum "workable" shares (fig. 5).

The physical results of subdivision on the geometry of urban fabric can be readily noticed. Initially, regardless of topographic constraints, no oblique partition lines are made in undeveloped buildable land. After construction, subdivision lines follow exterior walls and partitions that mostly intersect at right angles. Dead-end alleys emerge as an outcome of subdivision in response to the location of some shares deep within residential blocks away from streets. Over time, their number increases as a consequence of the number of subdivision operations. In most cases, courtyards tend to be square in shape and centrally located in the plot. Despite the intensive use of the courtyard house as a preferred typology, because it insures privacy and provides a good response to harsh climatic conditions, sharers tend to compete with each other to have portions of elevations on the streets. Consequently, resulting shares tend to be deep, with smaller elevations along streets.

Succession law has also had an impact on the subdivision of water, irrigation systems and water networks in desert cities. To a large extent, channels irrigating palm trees and gardens reflected the subdivision of water into shares. Delta elements were constructed as points of bifurcation for the diverging water channels [Bonine 1979; Ben-Hamouche 2008]. Their number along the path of water from the source to its destination reflected the stages of subdivisions. In case of conflicts of passages and crossings, canals were constructed like overpasses that could go as high as three levels. Tamentit, south of Algeria, provides an excellent example of such water subdivisions and the extent to which water systems shaped the landscape and the urban geometry of the city (fig. 6).

Postscript: Succession law and rrban regeneration policies

At present, most old Muslim cities are suffering from a continuous degradation due to various and interacting factors. The failure to absorb modernity and the overwhelming new ways of life, which are based on private cars, electronic devises such as washing machines, electricity, and gas, are considered determinant factors. The bright lights of big, new cities have also driven most of original populations out of the old cities, in search of a decent life. Consequently, this has turned the old cities into slums for poor classes of inland migrants. Politically, the interest of public authorities is mostly directed to new developments, which are used by the politicians in electoral campaigns as proof of achievement and signs of progress, which has heavily marginalized the development of the old cities.

Succession law itself, which for centuries was a source of urban dynamics, has in a way turned into a negative factor and a source of degradation. Its abolishment led to an institutional crisis. During the pre-colonial period, most cities were equipped with an institution for subdivision of inheritance. This was the case of the Ottoman institution called *shughl al mawarith al makhziniyya*, which was in charge of the liquidation of properties of deceased and lost persons. Subdivision was made as soon as a person was buried [Ben-Hamouche M. 2009b]. Properties were listed, sorted out, and then subdivided among heirs in the presence of a representative of the court. All too often the public treasury was the main heir for all the devolved assets during major catastrophes such as wars, droughts, and epidemics.

In the absence of these old institutions, most buildings and properties that socially fall into inheritance are witnessing a form of dispersed ownership and thus often becoming no man's land. Over decades of unsolved problems of inheritance, rights of generations of heirs have accrued over most of the old decaying buildings. Achieving an agreement among these generations became an impossible mission for co-owners, most of whom are either absent or disinterested. Tenants who are still present in the place make no effort to improve or maintain the old buildings, which have a low turnover and suffer from advanced states of decay. An initial step in a further study would have the goal of detecting the number of buildings in old cities that suffer from such a problem and investigating in the feasibility of re-inventing the institution of subdivision of inherited properties.

Conclusion

This study has highlighted the system of Islamic succession law and its direct impact on the physical environment. The morphology of old Muslim cities cannot be analyzed without understanding the mechanisms of such a system. Mathematics, sociology and engineering were combined under this system and generated the dynamics of these cities.

It is also evident that any preservation or regeneration policy aimed at improving the situation in old cities must necessarily pass though the reconsidering of the Islamic succession law and the re-establishment of the institutions in charge of its implementation.

Notes

1. In its origin it was authored by the jurist Siraj Al-deen Muhamed Al_sajawandi, who died around the 13th c. A.D., while Professor Almaric Rumsey (1825-1899), of King's College, London, was the author of many works on the subject of the Muslim law of inheritance and a barrister-at-law.
2. A false male ancestor is where a female ancestor intervenes in the line of descent.
3. It is the intention of the author in future research to automate the succession law and further highlight the impact on the physical environment.
4. Examples of such elements are given in B.S. Hakim [2009: 29-41].
5. Although it is possible to find real examples of subdivision, it is beyond the scope of this paper as it requires a social survey of heirs, a collection of legal records over decades, and a physical survey on the ground of the subdivided assets.

Bibliography

ABDUL-JAMAL, R. 2005. The Issue On Inheritance for Women available online: http://www.bismikaallahuma.org/archives/2005/the-issue-on-inheritance-for-women/. Last accessed 5 Sept. 2010.

AKBAR, J. 1988. *Crisis in the built environment*. Singapore: Concept Mass Media.

AMRANI (AL), M. 2000. *Al-Myrath*. Algiers: ANEP.

BARBIER, 1900. Droit Musulman: Des droits et obligations entre propriétaires. *Revue Algérienne et Tunisienne de Législation et de Jurisprudence* **XVI**: 9-114.

BATTY, M. 1991. Cities as Fractals: Simulating Growth and Form. In *Fractals and Chaos*, R. Crilly, A. Earnshaw and H. Jones, eds. New York: Springer-Verlag.

BECHHOEFER, W. and M. APPLEBY. 1997. Fractals, Music and Vernacular Architecture: An Experiment in Contextual Design Traditional Dwellings and Settlements. In *Critical Methodologies in the Study of Traditional Environments* **97**. Berkeley: University of California at Berkeley, Centre for Environmental Design.

BECHHOEFER, W. and C. BOVILL. 1994. Fractal Analysis of Traditional Housing in Amasya, Turkey. *Proceedings of the fourth Conference of the International Association for the Study of Traditional Environments*. Berkeley: International Association for the Study of the Traditional Environment.

BEN-HAMOUCHE, M. 2003. Decision-Making System and Urban Geometry: The Case of Algiers. *Journal of Architectural & Planning Research* **20**, 4: 307-322.

————. 2008. Islamic Law for Water and Land Management and its Impact on Urban Morphology. Second International Conference on Built Environment in Developing Countries, Penang, Malaysia, 1-3 December, 2008.

————. 2009a. Can Chaos Theory Explain Complexity In Urban Fabric? Applications in Traditional Muslim Settlements. *Nexus Network Journal* **11**, 2: 217-242.

————. 2009b. Complexity of urban fabric in traditional Muslim cities: Importing old wisdom to present cities. *URBAN DESIGN International* **14**, 1: 22-35.

————. 2009c. *Dar Es-Sultan L'Algérois à l'époque Ottomane.* Algiers: Dar El-Bassair.

BONINE, M. E. 1979. The Morphogenesis of Iranian Cities. *Annals of the Association of American Geographers* **69**, 2: 208-224.

BOVILL, C. 1996. *Fractal Geometry in Architecture and Design.* Boston: Birkhäuser.

BOVILL, C. 1996. *Fractal Calculations in Vernacular Design in Traditional Dwellings and Settlements.* International Association for the Study of Traditional Environments (IASTE) Working Paper Series **97**: 35-51. Berkeley: IASTE.

BRIGGS, J. 1992. *Fractals: The Patterns of Chaos: Discovering a New Aesthetic of Art, Science, and Nature.* New York: Touchstone.

BRUNSCHWIG R. 1947. Urbanisme médiéval et droit musulman. *Revue des Etudes Islamiques* **XV**: 127-155; rpt. in *Etudes d'islamologie* (Paris : Maisonneuve et Larose, 1976), vol. II, 7-35.

CA'FER EFENDI. 1987. *Risale-I Mimariyye: An Early Seventeenth Century Ottoman Treatise on Architecture.* Howard Crane, ed. and trans. Leiden: E. J. Brill.

CAMI-EFENDI. 2000. *Riyadh Al-Qasemeen.* M. Ben-Hamouche, ed. Damascus: Dar-Al-Bashair.

DE SOTO, Hernando 2000. *The Mystery of Capital: Why Capitalism Triumphs in the West and Fails Everywhere Else* New York: Basic Books.

DEVANEY, R. L. 1990. *Chaos, Fractals and Dynamics: Computer Experiments in Mathematics* Addison-Wesley Pub. Co.

EGLASH, R. 1999. *African Fractals: Modern Computing and Indigenous Design.* Piscataway NJ: Rutgers University Press.

HAKIM, B. S. 1988. *Arabic-Islamic Cities.* London and New York: Kegan Paul International.

————. 2009. *Sidi Boussa'id, Tunisia: Structure and Form of a Mediterranean Village.* Emergent City Press.

HAMILTON, C., trans. 1989. *The Hedaya : commentary on the Islamic laws.* Karachi: Darul Ishaat.

HUMA, A. 1997. *Muslim Contributions to Science, Philosophy, and the Arts* available online http://www.jannah.org/articles/contrib.html. Last accessed on Sept. 5th 2010.

LETOURNEAU, R. 1987. *Fès avant le protectorat : étude économique et sociale d'une ville de l'occident musulman* Publications de l'Institut des hautes études marocaines, t. 45. Rabat: Éditions La Porte.

OSTWALD, M. 2001. Fractal Architecture: Late Twentieth-Century Connections Between Architecture and Fractal Geometry. *Nexus Network Journal* **3**, 1: 73-84.

RAYMOND, A. 1994. Islamic City, Arab City: Orientalist Myths and Recent Views. *British Journal of Middle Eastern Studies* **21**, 1: 3-19.

————. 1985. *Les Grandes Villes Arabes a l'Epoque Ottomane.* Paris: Sindbad.

RUMSEY, A., ed. and W. JONES, trans. 1869. *Al-Sirajiyyah: or Mohammedan Law of Inheritance.* London: William Amer; rpt. Whitefish, MT: Kessenger Publishers, 2009.

SZEMPLINSKA-STUPNICKA, W. 2003. *Chaos, Bifurcations and Fractals Around Us: A Brief Introduction.* World Scientific Series on Nonlinear Science, Series A, vol. 47, Singapore: World Scientific Publishing Company.

UN-HABITAT. 2005. *Islam, Land & Property Research Series* Paper 6: Islamic Inheritance Laws & Systems. http://www.unhabitat.org/downloads/docs/3546_3490_ILP%206.doc. Last accessed 18 November 2010.

YEKINI, A. O. 2008. Women and Intestate Succession in Islamic Law. Lagos State University available at: http://ssrn.com/abstract=1278077. Last accessed on September 20, 2010.

About the author

Mustapha Ben-Hamouche is an associate professor at the University of Bahrain. where he teaches Islamic architecture and contemporary issues in Arab Muslim cities. He previously worked as an expert planner in the department of town planning in Al-Ain City, United Arab Emirates. He is interested in the application of chaos theory and fractals to Muslim cities, geographic information systems (GIS) and cellular automata in urban planning.

Liliana Curcio Roberto Di Martino
LA PROSPETTIVA A 180° E OLTRE
Il punto di vista di Emilio Frisia

Keywords: Liliana Curcio,
Roberto Di Martino,
perspective, Emilio Frisia

Book Review

Liliana Curcio and Roberto Di Martino

La prospettiva a 180° e oltre. Il punto di vista di Emilio Frisia

Quaderni del Centro Internazionale Urbino e la Prospettiva 4

Urbino: Centro Internazionale di Studi "Urbino e la prospettiva", 2009

Reviewed by Sylvie Duvernoy

Via Benozzo Gozzoli, 26
50124 Florence ITALY
syld@kimwilliamsbooks.com

Emilio Frisia (1924-2004), an Italian professional photographer, was the son of Donato Frisia, a painter acquainted with Modigliani and Guttuso. Frisia taught drawing and descriptive geometry at the Istituto Statale d'Arte of Monza, near Milan, for many years. His artistic talent, coupled with his scientific skills, made him an extraordinary teacher. His passion for all kinds of graphic art, whether achieved with mechanical tools (photographic device or computer technology) or manual means (pencils, watercolour, etc.) led him to investigate in depth all the opportunities offered by the combination of modern and traditional graphic tools, in order to produce fantastic views of imaginary spaces.

Authors Liliana Curcio and Roberto di Martino, who are both teachers at the same school where Frisia used to work, present the results of his research. The authors are well-known to the readers of the *Nexus Network Journal* and followers of the Nexus conference; they organized the exhibits of student work that accompanied Nexus 2000 in Ferrara and Nexus 2006 in Genoa.

The artworks by Emilio Frisia published in this book are all two-dimensional printed images, rendered by hand, representing unnatural three-dimensional views of virtual outdoor or indoor spaces, to which the author gives poetic titles such as "Staircase on the seashore, climbing towards the unknown" or "Solitude of a couple in the museum".

In order to produce his three-dimensional views Frisia makes use of all kinds of geometric projection: linear perspective, axonometric projection, cartographic projections, and more. Sometimes more than one step is necessary to construct the final view. Furthermore, effects of reflections on mirroring surfaces, and/or effects of various light refractions are added to the imaginary space to evoke some particular feelings in the viewer; this is the case, for instance, of the picture entitled "nightmares in a glass cube".

A series of images entitled "Una città ideale – an ideal city", dedicated to Piero della Francesca, show a urban space which is successively represented thanks to different kinds of map projections, such as stereographic projection, Mercator projection, Sanson-Flamsteed projection, etc.

DOI 10.1007/s00004-011-0063-7; *published online* 7 April 2011

Even though Liliana Curcio and Roberto di Martino did not intend the book as a textbook meant to explain how to build sophisticated three-dimensional views, it clearly points out the relationships existing between the major conic projections, such as linear perspective, anamorphosis and cartography, which traditional teaching does not always sufficiently explain. The sheer beauty of the illustrations is enough to explain why modern students should continue, even today, to study and learn the techniques of map projection. In fact, cartography is still taught to architecture students, at least in Italy, although the drawing of maps has nowadays become a highly specialized task, evolving together with modern computer and geospatial technologies, which contemporary architects are not expected to master. The didactic purpose of studying map projection lies in the observation of the close, reciprocal connection that exists between representation techniques and final goals. Moreover, the possible applications of techniques of map projection and spherical perspective are not restricted to cartography, but can be used to construct to views of complex curved architectural forms, for which the usual methods of orthographic projection (section and/or elevation), or of linear perspective, do not result in images that are sufficiently expressive and communicative.

Emilio Frisia is obviously not an isolated exception in the wide world of drawing teachers. Many other talented artists are currently teaching geometry and drawing to art and architecture students around the world, and many of these produce valuable works of art. But the collection of Frisia's beautiful drawings presented here show that even artwork produced in the realm of didactic activity and didactic experimentation may deserve to be published, simply for the sake of being seen and admired.

About the reviewer

Sylvie Duvernoy is Book Review Editor of the *Nexus Network Journal.*

NEXUS NETWORK JOURNAL   Architecture and Mathematics

Subscription information

ISSN print edition 1590-5896
ISSN electronic edition 1522-4600

Subscription rates

For information on subscription rates please contact:
Springer Customer Service Center GmbH
The Americas (North, South, Central America and the Caribbean)
journals-ny@springer.com
Outside the Americas: subscriptions@springer.com

Orders and inquiries

The Americas (North, South, Central America and the Caribbean)
Springer Journal Fulfillment
P.O. Box 2485, Secaucus
NJ 07096-2485, USA
Tel.: 800-SPRINGER (777-4643)
Tel.: +1-201-348-4033 (outside US and Canada)
Fax: +1-201-348-4505
e-mail: journals-ny@springer.com

Outside the Americas

via a bookseller or
Springer Customer Service Center GmbH
Haberstrasse 7, 69126 Heidelberg, Germany
Tel.: +49-6221-345-4304
Fax: +49-6221-345-4229
e-mail: subscriptions@springer.com
Business hours: Monday to Friday
8 a.m. to 8 p.m. local time and on German public holidays

Cancellations must be received by September 30 to take effect at the end of the same year.

Changes of address: Allow six weeks for all changes to become effective. All communications should include both old and new addresses (with postal codes) and should be accompanied by a mailing label from a recent issue.

According to § 4 Sect. 3 of the German Postal Services Data Protection Regulations, if a subscriber's address changes, the German Post Office can inform the publisher of the new address even if the subscriber has not submitted a formal application for mail to be forwarded. Subscribers not in agreement with this procedure may send a written complaint to Customer Service Journals,within 14 days of publication of this issue.

Back volumes: Prices are available on request.

Microform editions are available from: ProQuest. Further information available at: http://www.il.proquest.com/umi/

Electronic edition

An electronic edition of this journal is available at springerlink.com

Advertising

Ms Raina Chandler
Springer, Tiergartenstraße 17
69121 Heidelberg, Germany
Tel.: +49-62 21-4 87 8443
Fax: +49-62 21-4 87 68443
springer.com/wikom
e-mail: raina.chandler@springer.com

Instructions for authors

Instructions for authors can now be found on the journal's website: birkhauser-science.com/NNJ

Production

Springer, Petra Meyer-vom Hagen
Journal Production, Postfach 105280,
69042 Heidelberg, Germany
Fax: +49-6221-487 68239
e-mail: petra.meyervomhagen@springer.com
Typesetter: Scientific Publishing Services (Pvt.) Limited, Chennai, India
Springer is a part of
Springer Science+Business Media
springer.com
Ownership and Copyright
© Kim Williams Books 2011